Planning and Engineering Guidelines for the Seismic Retrofitting of Historic Adobe Structures

Planning and Engineering Guidelines for the Seismic Retrofitting of Historic Adobe Structures

E. Leroy Tolles

Edna E. Kimbro

William S. Ginell

The Getty Conservation Institute   Los Angeles

Getty Conservation Institute
Scientific Reports series

© 2002 J. Paul Getty Trust

Getty Publications
1200 Getty Center Drive, Suite 500
Los Angeles, California 90049-1682
www.getty.edu

Timothy P. Whalen, *Director, Getty Conservation Institute*
Jeanne Marie Teutonico, *Associate Director, Field Projects and Conservation Science*

Dinah Berland, *Project Manager*
Leslie Tilley, *Manuscript Editor*
Pamela Heath, *Production Coordinator*
Garland Kirkpatrick, *Cover Designer*
Hespenheide Design, *Designer*

Printed in the United States of America

The Getty Conservation Institute works internationally to advance conservation and to enhance and encourage the preservation and understanding of the visual arts in all of their dimensions—objects, collections, architecture, and sites. The Institute serves the conservation community through scientific research, education and training, field projects, and the dissemination of the results of both its work and the work of others in the field. In all its endeavors, the Institute is committed to addressing unanswered questions and promoting the highest possible standards of conservation practice.

The Getty Conservation Institute is a program of the J. Paul Getty Trust, an international cultural and philanthropic organization devoted to the visual arts and the humanities that includes an art museum as well as programs for education, scholarship, and conservation.

The GCI Scientific Program Reports series presents current research being conducted under the auspices of the Getty Conservation Institute. Related books in this series include *Seismic Stabilization of Historic Adobe Structures: Final Report of the Getty Seismic Adobe Project* (2000) and *Survey of Damage to Historic Adobe Buildings After the January 1994 Northridge Earthquake* (1996).

All photographs are by the authors unless otherwise indicated.

Library of Congress Cataloging-in-Publication Data
Tolles, E. Leroy, 1954–
  Planning and engineering guidelines for the seismic retrofitting of
historic adobe structures / E. Leroy Tolles, Edna E. Kimbro,
William S. Ginell.
     p. cm. — (GCI scientific program reports)

  ISBN 0-89236-588-9 (pbk.)

  1. Building, Adobe. 2. Buildings—Earthquake effects. 3. Earthquake
engineering. 4. Historic Buildings—Southwest (U.S.)—Protection.
I. Kimbro, Edna E. II. Ginell, William S. III. Title. IV. Series.
  TH4818.A3 T65 2003
  693'.22—dc21
                                                  2002013971

Contents

Foreword

Adobe, or mud brick, is one of the oldest and most ubiquitous of building materials. Extremely well suited to the hot, dry climates and treeless landscapes of the American Southwest, adobe became a predominant building material in both the deserts of New Mexico and Arizona and throughout early California. The vulnerability of many original adobe structures to damage or destruction from earthquakes has been of serious concern to those responsible for safeguarding our cultural heritage.

The guidelines presented here address the practical aspects of this problem and represent the culmination of twelve years of research and testing by the Getty Conservation Institute and its partners on the seismic retrofitting of adobe buildings. The Getty Seismic Adobe Project (GSAP) was first discussed in October of 1990 at the Sixth International Conference on Earthen Architecture in Las Cruces, New Mexico. This is the third in a series of GSAP volumes that began in 1996 with the publication of the *Survey of Damage to Historic Adobe Buildings After the January 1994 Northridge Earthquake* and continued in 2000 with the publication of *Seismic Stabilization of Historic Adobe Structures*.

The Getty Seismic Adobe Project and its reports manifest the GCI's commitment to identify and evaluate seismic retrofitting methodologies that balance safety with the conservation of our cultural heritage. To this end, the overarching objective of GSAP was to find technologically feasible, minimally invasive, and inexpensive techniques with which to stabilize adobe buildings. The GSAP team conducted research and testing of adobe structures to evaluate retrofitting methodologies that would ensure adherence to safety standards while preserving the historic architectural fabric. This publication is, in a very real sense, the end product of that project.

These guidelines can assist responsible parties in the planning of seismic retrofitting projects that are consistent with both conservation principles and established public policy; they can help local officials establish parameters for evaluating submitted retrofitting proposals; and they can serve as a resource for technical information and issues to be considered in the design of structural modifications to historic adobe buildings.

Planning and Engineering Guidelines for the Seismic Retrofitting of Historic Adobe Structures examines the two phases necessary to improving the performance of an adobe building during an earthquake; both planning and engineering design are critical to the success of such a project, and both are described in detail here. In the appendixes, relevant governmental standards are reprinted and sources of further information suggested. To make this information as accessible as possible, a cumulative index to all three volumes in the GSAP series has been provided.

The major credit for this research and its practical results goes to William S. Ginell, senior scientist at the GCI, who has been the driving force behind both GSAP and this series of publications. Without Bill's consistent and dedicated leadership on the project, it would not have become the significant contribution to the field that it is. I am grateful to him for the knowledge, experience, and unwavering commitment he has brought to this endeavor.

I also want to thank E. Leroy Tolles and Edna E. Kimbro for their dedicated work on GSAP over the years. Their efforts have resulted in a body of work that is useful, comprehensible, and a cornerstone in the field of seismic retrofitting studies. Both have contributed immeasurably to the project and to this final volume in the series.

It is my hope that this book will be of use to all those charged with preserving our earthen architectural heritage in seismically active regions.

Timothy P. Whalen
Director, The Getty Conservation Institute

Acknowledgments

In compiling these guidelines for planning and implementing the conservation of historic adobe structures, the authors considered it important to solicit and consider the responses of the many colleagues who had been faced with similar problems to those being discussed. Their experiences, often based on differing philosophies and approaches to the issues, enriched our understanding of the many aspects of seismic stabilization that needed to be considered when contemplating changes to culturally significant buildings. Although their views differed occasionally in some respects from our own, their opinions were nonetheless appreciated and we are grateful for their contributions.

The authors would like to thank Neville Agnew and Frank Preusser, who initiated, encouraged, and supported the Getty Seismic Adobe Project (GSAP) at The Getty Conservation Institute. This book represents the culminating document of the work they helped to formulate beginning in 1990. We would like to acknowledge the important contributions to the GSAP made by Charles C. Theil Jr. and Frederick A. Webster, who were members of the research team throughout much of the planning and experimental studies of the seismic behavior of model adobe structures. We also appreciate the efforts of the GSAP Advisory Committee members who provided oversight of the project's research, and those of Helmut Krawinkler and Anne Kiremidjian of Stanford University and Predrag Gavrilovic and Veronika Sendova of IZIIS, Republic of Macedonia, who supported our efforts during the shaking-table tests of model adobe structures at their institutions.

Tony Crosby, Wayne Donaldson, John Fidler, and Nels Roselund read parts of the manuscript at various times during its development. They asked searching questions and provided comments on the planning and engineering aspects of the drafts that helped to make this book more useful. We are also indebted to two anonymous peer reviewers who astutely commented on the text and suggested numerous ways in which the guidelines could be improved and made clearer.

Thanks are also due to Gary Mattison at the GCI, who assembled the manuscript and who shepherded it through innumerable drafts, and to the Getty Publications staff and its consultants, who brought the book to light: Dinah Berland, project manager; Leslie Tilly, manuscript editor; Pamela Heath, production coordinator; and Gary Hespenheide, designer.

Introduction

Many of the early structures in the southwestern United States were built of adobe. The materials available for construction of early churches, missions, fortifications (presidios), stores, and homes were generally limited to those that were readily available and easily worked by local artisans. Adobe has many favorable characteristics for construction of buildings in arid regions: it provides effective thermal insulation, the clayey soil from which adobe bricks are made is ubiquitous, the skill and experience required for building adobe structures is minimal, and construction does not require the use of scarce fuel. As a consequence of their age, design, and the functions they performed, surviving historic adobe structures are among the most historically and culturally significant structures in their communities.

However, earthquakes pose a very real threat to the continued existence of adobe buildings because the seismic behavior of mudbrick structures, as well as that of stone and other forms of unreinforced masonry, is usually characterized by sudden and dramatic collapse. There is also the threat to occupants and the public of serious physical injury or loss of life during and following seismic events. Generally speaking, it is the evaluation of the engineering community that adobe buildings, as a class, are more highly susceptible to earthquake damage than are the various other types of buildings.

Nevertheless, it has been observed that some unmodified adobe buildings have withstood repeated severe earthquake ground motions without total collapse. On this point, a prominent seismic structural engineer remarked, "The common belief that a building is strong because it has already survived several earthquakes is as mistaken as assuming that a patient is healthy because he has survived several heart attacks" (Vargas-Neumann 1984).

The seismic upgrading of historic buildings embraces two distinct and apparently conflicting goals:

- Seismic retrofitting to provide adequate life-safety protection
- Preservation of the historic (architectural) fabric of the building

These goals are often perceived as being fundamentally opposed.

If conventional seismic retrofitting practices are followed, extensive alterations of structures are usually required. These alterations can involve the installation of new structural systems and often substantial removal and replacement of existing building materials. However, historic structures so strengthened and fundamentally altered may lose much of their authenticity. They are virtually destroyed by the effort to protect against earthquake damage, before an earthquake even occurs. Thus, the conflict is seen to be between retrofitting an adobe building to make it safe during seismic events, at the cost of destroying much of its historic fabric in the process, and keeping the historic fabric of the building intact but risking structural failure and collapse during future seismic events.

Faced with the apparent standoff between unacceptable seismic hazard and unacceptable consequences of conventional retrofitting approaches, the Getty Conservation Institute has made a serious commitment to identify and evaluate seismic retrofitting methodologies for historic adobe structures that balance the need for public safety with the conservation of cultural assets. The Getty Conservation Institute's Seismic Adobe Project (GSAP) is the manifestation of that commitment.

The GSAP, of which these guidelines are a product, was directed toward the development of seismic-retrofitting technology for the protection of culturally significant earthen buildings in seismic areas. Seismic retrofitting is the most viable means of minimizing catastrophic earthquake damage to cultural monuments and simultaneously assuring the safety of building occupants. The objective of the project was to develop methods for safeguarding the authenticity of earthen architectural structures by preservation of culturally significant building fabric while protecting the buildings—and consequently the public—from drastic, often catastrophic, earthquake effects.

The response of retrofitted buildings to several recent earthquakes has amply demonstrated that appropriate modification of buildings before a seismic event is a practical solution. A survey of historic adobe buildings damaged in the 1994 Northridge earthquake showed that retrofitted adobe buildings performed significantly better than those that had not been upgraded. However, damage repair and conventional retrofitting after a major earthquake can be very costly. It has even led, in some notable instances, to demolition of important landmarks for primarily financial reasons.

The Purpose of These Guidelines

In an address to an international conference devoted to seismic strengthening of cultural monuments, Jukka Jokilehto said, "The principle [of historic architecture conservation] is to define the architectural typology of the buildings, to define the historical parameters in deciding what is the substance that conserves the historic value of the fabric, and where changes can be proposed. The principle is to define the limits of transformability, and prepare positive guidelines for the owners and users of the buildings to develop their properties in harmony with the principles of preservation of urban historic values" (Jokilehto 1988:7).

The primary objective of these guidelines is to assist owners, cultural resource managers, and other responsible parties in "defining the limits of transformability" and in planning seismic retrofit projects consistent with conservation principles. The guidelines are intended to assist nonprofit organizations, churches, private owners, and government agencies that follow established public policy (such as provisions of the National Environmental Protection Act, the California Environmental Quality Act, and Public Resources Code sections 5024 and 5024.5) regarding cultural resource protection.

A second objective of these guidelines is to provide information to help local historic resource commissions and officials responsible for historic resource protection evaluate technical retrofit proposals submitted for review. In the past, historical resource commissions charged with reviewing alteration permits for historic structures have experienced some difficulty in evaluating the responsiveness of applications for seismic retrofitting to accepted preservation standards. Proposals before historic resource commissions must be evaluated according to the specific criteria contained in the preservation ordinances of each jurisdiction.

A final purpose of these guidelines is to present information on factors to be considered when planning the restoration of a historic adobe structure, where further information can be obtained, and what technical information is available that can be used in the design of structural modifications of historic adobe buildings. The information presented here has been obtained from the published literature, observations made during surveys of adobe buildings, experimental studies of the behavior of adobe structures during simulated seismic excitation, analyses of representative adobe retrofit projects, and prevailing building standards.

About This Book

The Getty Seismic Adobe Project was established to develop and test alternative retrofitting techniques that are inexpensive, easily implemented, and less invasive than those currently in use (see appendix A). The long-term goal of GSAP was to develop design practices and tools that would be tested, documented, and made available for distribution to the interested public. Recognizing that planning and implementation are separate tasks, and that the two require coordination, this book addresses this goal and describes the two phases of a program designed to improve the seismic performance of historic adobe structures: a planning phase and an engineering-design phase.

The techniques described may be of interest to

- owners, managers, and government agencies concerned with historic properties;

- government officials responsible for the formulation and implementation of building codes and public safety regulations;
- architects;
- engineers;
- architectural historians;
- architectural conservators;
- historic structure preservation contractors; and
- public groups committed to the preservation of our historic patrimony.

The engineering information presented here is based on the experience gained from extensive studies of early work on seismic stabilization, on recent shaking-table tests on model adobe structures (Tolles et al. 2000), and on the results of in situ evaluation of the damaging effects on adobe structures of the 1994 Northridge, 1989 Loma Prieta, 1987 Whittier Narrows, and 1971 San Fernando (Sylmar) earthquakes in California (Tolles et al. 1996).

The guidelines are intended for use wherever historic adobe buildings are threatened by seismic activity. Anecdotal information drawn from experience at various historic adobe sites in California is included in order to relate conservation theory to actual preservation practice. Some of the material is specific to California law; however, many communities around the world that are as seismically active as those in California have enacted, or contemplate passage of, ordinances relating to seismic retrofitting. In the United States, the majority of historic adobe buildings at greatest risk of earthquake damage are located along the California coast in Seismic Zone 4, where Spanish colonization efforts were concentrated. Thus, the problems in California represent case studies of issues and concerns that are widespread throughout both the New and Old Worlds.

We believe that the results of our research should be widely applicable and that the general planning methodologies outlined in these guidelines are valid in all regions of high seismicity. Virtually all states in the United States have offices of historic preservation that are responsible for inventorying and overseeing the conservation of cultural landmarks, and similar agencies exist in other nations as well. Standards and charters, such as the Venice and Burra Charters, are widely recognized and applied throughout the world (ICOMOS 1999). Although specific conditions, professional services, and other factors can vary locally and may differ from those in California, the general ideas presented in these guidelines should be applicable elsewhere as well. Those responsible for the maintenance and preservation of historic adobe buildings are urged to begin planning seismic retrofit programs for the buildings in their care.

Chapter 1

Conservation Issues and Principles

The loss of California's historic adobe architecture began with widespread earthquake damage to homes and the Spanish missions in the early 1800s, and the threat continues largely unabated today. Surveys indicate that most historic adobe buildings have not been strengthened to withstand seismic events and that many have already been weakened during earthquakes. Yet time and again the world over, seismically retrofitted buildings have survived earthquakes, sustaining only reparable damage, while ancient unstrengthened buildings have been lost. For example, strengthened Spanish Colonial buildings in Cuzco, Peru, survive despite frequent seismic activity, and Mission Dolores, strengthened following the 1906 San Francisco earthquake (fig. 1.1a), performed well in the 1989 Loma Prieta earthquake (fig. 1.1b).

In California, organized efforts to preserve adobe architectural landmarks began before the turn of the twentieth century with the precursor of Charles Lummis's Landmarks Club. These efforts were followed by those of Joseph Knowland and the California Historic Landmarks League. The Landmarks Club was instrumental in preserving missions in Southern California, while the Landmarks League was active in the north in the preservation of Old Town Monterey and Mission San Antonio de Padua in the early 1900s. The Native Sons and Daughters of the Golden West actively supported these efforts. The movement gained momentum with the reconstruction of Mission San Luis Rey by the Franciscans in the 1890s, San Miguel in the 1930s, and San Antonio in

Figure 1.1

Mission San Francisco de Asis (Dolores), San Francisco, Calif.: (a) after the 1906 San Francisco earthquake (courtesy Santa Barbara Mission archives), and (b) after the 1989 Loma Prieta earthquake.

(a)

(b)

the 1940s, the latter with Hearst family financial assistance. In the 1930s, the historic adobe buildings of the Spanish capital at Monterey were preserved by the Monterey History and Art Association and the California Division of Beaches and Parks (now California State Parks). Missions and other structures in San Juan Bautista, Sonoma, Santa Barbara, and San Diego were also upgraded and preserved.

In recent years, the toll on historic California adobes that did not receive such preservation efforts has been dramatic. Recent losses to unstrengthened buildings include destruction of the San Fernando Mission Church in 1971 (fig. 1.2); heavy damage to the Pio Pico Mansion in 1987 (fig. 1.3); damage to the Rancho San Andres Castro Adobe and Juana Briones Adobe in 1989; and severe damage to the Del Valle Adobe at Rancho Camulos (fig. 1.4), the Andres Pico Adobe (fig. 1.5), the De la Ossa Adobe (fig. 1.6), and the *conventos* of the San Gabriel and San Fernando missions in 1994.

Preservationists in the twenty-first century should continue to initiate and bolster preservation efforts by confronting, not backing

Figure 1.2
San Fernando Mission Church, Mission Hills, Calif.: (a) interior after the 1971 Sylmar earthquake (courtesy estate of Norman Neuerburg), and (b) exterior demolition after the 1971 earthquake (courtesy San Fernando Valley Historical Association).

Figure 1.3
Two exterior views of Pio Pico Mansion, Whittier, Calif., after the 1987 Whittier Narrows earthquake.

(a)   (b)

(a)   (b)

away from, the challenge that earthquakes pose to many of the state's earliest cultural resources—the missions and other historic adobe buildings. In California, preservation of the state's Spanish Colonial and Mexican-era adobe buildings is an important concern, not only because

Figure 1.4
Rancho Camulos Museum, Piru, Calif., after 1994 Northridge earthquake: (a) collapse of southwest corner bedroom, and (b) damage at top of wall.

(a)   (b)

Figure 1.5
Andres Pico Adobe, Mission Hills, Calif.: (a) before 1994 earthquake, and (b) after 1994 earthquake.

(a)   (b)

Figure 1.6
De la Ossa Adobe, Encino, Calif.: (a) before 1994 Northridge earthquake, and (b) after 1994 Northridge earthquake.

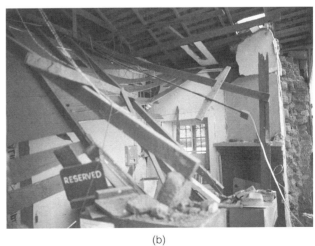

(a)   (b)

of their vulnerability to earthquake damage but also because of their growing social, historical, and ethnic cultural value in a society that is becoming increasingly Hispanic in cultural orientation.

The Significance of Adobe Architecture in California

Historic adobe buildings in California are defined largely by their material and are often named for it: the Los Coches Adobe, the Rancho San Andres Castro Adobe, the Las Cruces Adobe. Both the historical and artistic or architectural qualities of adobe structures can be attributed in whole or part to their materials. The material itself and the traditional ways of building with adobe contribute a great deal to the aesthetic qualities of historic adobe structures.

The historic fabric of adobe buildings—including the raw building materials as altered and assembled to form buildings: the forming of the bricks, the harvesting of the timber, the working of the wood—represents modifications of natural materials by human labor, which impart to the resulting buildings an additional dimension that is worthy of respect: the "handiwork of past artisans," to quote the Venice Charter (Riegl 1964).

Mud-brick walls, lintels, *vigas* (beams), roof framing, tiles, and original renderings—particularly if painted, incised, or decorated— possess significance, both in their appearance and materials. In wall paintings, not only the design but also the paint itself is important, the latter in that it can provide information about the past use of materials. Similarly, the plant content of adobe bricks can be analyzed to identify the plant species used as reinforcing material and to differentiate between native plants and those imported from elsewhere. Significance lies not only in the aesthetic form but also in the substance of the materials.

The mud bricks of historic adobe monuments can be compared to the "stones of Venice," which the nineteenth-century art and architecture critic John Ruskin (1819–1900) viewed as living testimonies to the work of past generations (Ruskin 1851–53). Fabricated assemblages, both bricks and buildings, are products of human activity, possessing inherent integrity and meriting respect. Generally considered an archaic building material in California today, adobe is associated with the culture of the Native American and Hispanic ethnic groups that initially inhabited and settled the Spanish borderlands.

To better understand the cultural significance of historic adobe fabric in California, some observations about the state's Spanish Colonial architectural history are relevant. In California, the last of Spain's New World ventures, adobe buildings were handmade by the indigenous people. The adobe buildings they erected represent the largest and most visible products of their industry. Foundation stone was quarried and dressed by hand. Wood timbers were harvested, sawn, hewn, and finished. Earth and straw were mixed and formed, dried, turned, and stacked. Clay for roof and floor tiles was mined, formed, and baked. Limestone was quarried, burned, and slaked. And all this work was done by Native Americans trained by Spanish soldiers, engineers, clergy, and artisans. In the rancho era following the mission period, former mission neophytes made up the trained labor force employed to construct adobe buildings for the landed rancheros. Thus, practically speaking, one goal of conservation efforts is to preserve the physical results of the training in building skills received by the Native Americans and the manifestations of their contributions that are visible today in the missions and historic adobe buildings of early California.

Vestiges of neophyte craftsmanship are rare in California, and the critical Native American role in building the state's historical adobe structures has been recognized only relatively recently (Thomas 1991). Adobe buildings, sacred and secular, are arguably among the most tangible reminders of the work and lives of these people. As awareness grows about the role of Native Americans in the building of early California, adobe buildings will be increasingly valued for themselves, as well as for their long-appreciated Hispanic architectural and historical associations (fig. 1.7).

In New Mexico and Arizona, traditional adobe building practices, never truly extinct, are experiencing a revival. Such continuity with historic building traditions greatly facilitates maintenance and repair of adobe monuments. In California, however, the break with the past has, unfortunately, been almost complete. Ethnohistorical research suggests that the original craftsmen and builders have few identifiable descendants.

Figure 1.7
Examples of historic adobe buildings built with indigenous labor: (a) recently restored neophyte quarters at Mission Santa Cruz, Santa Cruz, Calif. (courtesy State Museum Resource Center, Calif. State Parks); (b) exterior of Mission La Purisima Concepcion, Lompoc, Calif.; and (c) *convento* of Mission La Purisima Concepcion.

(a)

(b)

(c)

The cultures that produced the architecture are not active in the field of contemporary adobe production, and contemporary building technology bears little relation to the historical methods described. Therefore, conservation of the remaining historic fabric, a process that is complicated by seismic events, must be assigned a high priority.

Historic adobe buildings in high-seismic-risk zones have almost invariably suffered previous damage and have been repaired repeatedly over the years. Cosmetic repairs may mask this reality, but historical photographs and careful investigation can substantiate this observation. After the 1989 Loma Prieta earthquake, the Rancho San Andres Castro Adobe exhibited areas where large sections of concrete had been used previously to fill wall cracks (fig. 1.8). The necessity of repairing cracks, replastering, and sometimes rebuilding sections of walls has traditionally been accepted in the cultures that produced these structures. However, when materials that are incompatible with adobe are used, such as concrete, the vulnerability to seismic damage is heightened. Accepting the inevitability of cracking and making timely repairs as a part of cyclical maintenance must continue, or the level of intervention necessary to repair cracking rises proportionately. This can result in substantial alterations to the building. Strengthening sufficient to minimize cracking requires interventions that result in considerable loss of historic fabric and a consequent loss of authenticity. For structures with highly significant surface features, the levels of acceptable intervention may have to be greater or different to minimize cracking and surface loss compared with less elaborate adobe structure surfaces that are cyclically renewed by traditional lime or mud-plaster treatments.

As concerned individuals contemplate the vulnerability of adobe landmarks to earthquakes and survey the damage already done, the need to redouble efforts to conserve such authenticity as exists and to use all the resources that conservation technology has to offer becomes apparent.

Figure 1.8
Rancho San Andres Castro Adobe, Watsonville, Calif., showing failure of early concrete repairs to adobe after the 1989 Loma Prieta earthquake.

Principles of Architectural Conservation

Three basic conservation principles guide the design of interventions contemplated for historically or culturally significant structures, regardless of the material or the location of the buildings:

1. Understanding the building
2. Minimal intervention
3. Reversibility

These general principles jelled over time and reflect efforts to reconcile two opposing approaches to the preservation of culturally significant buildings. Several conservation theorists contributed to resolving the contradictions between the nineteenth-century restoration/reintegration precepts of Viollet-le-Duc in France ([1858–72] 1959) and the stabilization/preservation ideals of Ruskin and his followers in England. In the twentieth century, a merging of philosophy and practice was achieved by applying traditional art conservation principles and methodology to structure-preservation efforts, resulting in the current analytical approach. The same architectural conservation principles that guided the treatment of the cathedrals of Notre Dame, Chartres, St. Paul's, and St. Peter's apply to the historic adobe buildings of the New World, including the Spanish missions and other historic adobe structures.

The first of the fundamental principles directs the conservator to know the building—to study and understand its materials, systems, condition, cultural and physical context, and history and alterations—before planning any intervention or alterations. The building is examined as a whole, in context, and each of its parts individually assessed. It is evaluated for aesthetic merit, unity or disunity as a formal work of architecture, and its historical value as a document of the passage of time or the flow of history. A historic structure report typically documents this analysis (see chap. 2 and appendix D). Any perceived conflicts among these fundamental values are resolved at this juncture through the evaluation of all available information and the contextual weighing and balancing of evidence to reach reasoned, justified, and nonarbitrary decisions. As a part of this information-gathering and analysis process, the need for alterations to the structure is assessed, which may be anything from replacement of a broken windowpane to seismic retrofitting.

Once the need for alteration has been identified, the second principle, that of minimal intervention, comes into play, influencing the design and extent of changes to the structure. Interventions or alterations are minimized to preserve as much of the significant fabric of the building as possible, thereby safeguarding its authenticity while accomplishing whatever goal motivated the initial decision to make alterations.

The third guiding principle, reversibility, holds that the alterations made to the building should be able to be removed in the future without significant damage to the building. Reversibility allows for the use of improved technologies as they are developed and the removal of inappropriate alterations. This principle encourages alterations of an additive nature and discourages the removal of material or architectural features. In addition, the permanent storage of any removed material or feature is important, to provide the opportunity for future replacement. Since all alterations become part of the history of the building, failure to achieve complete removability of the intervention results in permanent alterations.

However, complete eradication of an alteration is generally not feasible, and thus the concept of retreatability is gaining acceptance. *Retreatability* refers to the ability to alter a structure without the accumulation of visible, permanent changes, residues, or other negative consequences that could restrict future alterations (Oddy and Carroll 1999). In many instances, the goal of complete eradication of changes brought about by alteration cannot be achieved, and therefore retreatability—in the event that more appropriate materials or retrofitting procedures are developed later—is a more practical goal.

Other important tenets of contemporary architectural conservation include encouraging the use of identical, or similar but compatible, materials in the retrofitting and repair or replacement of deteriorated features to obtain similarity of performance. Acknowledging the necessity of ongoing maintenance to minimize the need for future interventions is also recognized as vital.

Seismic Retrofitting Issues

As stated at the beginning of the chapter, natural disasters, combined with rehabilitation and adaptive reuse, have taken a serious toll on California's earliest historic adobe structures. Of the more than 700 adobe structures originally constructed in the San Francisco Bay area during the Spanish and Mexican periods (Bowman 1951), perhaps 40 survive today, and the 1989 Loma Prieta earthquake further endangered some of these. An estimated 350 historic adobe structures (of varying degrees of integrity) dating from the Spanish and Mexican eras remain in the entire state of California (Thiel et al. 1991).

The state itself is responsible for the preservation of about seventy historic adobe structures; the others are under the jurisdiction of various public and nonprofit agencies, private individuals, and churches. Every historic adobe building in California is one of a steadily diminishing number of historic structures that are vulnerable to development pressures and earthquake damage, the two factors most often cited as being responsible for their demolition. It is therefore imperative that the remaining few be protected not only from the risk of damage or destruction by seismic events but also from the certain hazard of inappropriate countermeasures.

Life-Safety Issues

A fundamental goal of building regulations is to provide for adequate life safety during the largest seismic events. In California, the Unreinforced Masonry Building Law of 1986 (appendix B) requires that publicly accessible unreinforced masonry buildings (URM), of which adobe is one type, in Seismic Zone 4 be identified and mitigation programs prepared. After the 1989 Loma Prieta earthquake, some owners who had failed to retrofit their URM buildings were sued for damages for injury or loss of life caused by building collapse. Sound risk management dictates that such structures be retrofitted, demolished, or made inaccessible to the public.

A building that poses little life-loss hazard to its occupants can be judged a success even in the case of total economic loss due to earthquake damage. The intent of modern building codes is to specify minimal design features for preventing structural damage during moderate to major earthquakes, but these features might allow structural damage to occur during the most intense seismic events. Except for the most important facilities, buildings are designed with the assumption that structural damage can occur during very severe earthquakes.

Thus the first objective of seismic retrofitting measures is to minimize the potential for loss of life. Cracks in the walls and even structural damage may occur, but it is essential to provide for public safety by preventing structural instability and other risk factors for injury or loss of life. Seismic retrofit measures must first minimize the risk to lives and then satisfy the conservation principles of minimal intervention and reversibility. Only when all those criteria have been met should retrofit measures address the issue of minimizing damage to the building during moderate and major earthquakes. For example, measures that provide for life safety might have little effect on cracking during a moderate earthquake and could allow significant and nonreparable damage to occur during a major seismic event.

Conservation Issues

The importance of retaining the historic fabric of an adobe structure varies with each specific building and depends on what type of treatment is appropriate for that building: stabilization, preservation, restoration, rehabilitation, or reconstruction. A strict conservation approach is concerned with retention of the historic fabric and stylistic features above all else. However, to be able to specify the minimum intervention necessary and provide material compatibility and reversibility where possible, it is important to understand the adobe structure in the context of its history and environment prior to intervention. This first principle of conservation requires a broad, multidisciplinary assessment of the structure and identification of all its cultural values and historic fabric at varying levels of significance. These data are usually derived from a historic structure report (see chap. 2 and appendix D). In the seismic context, this information includes studies that document past seismic performance, microzonation, and results of geotechnical site investigations. Once such data are in hand, the principle of minimal intervention dictates that the least amount of alteration necessary to accomplish the task of seismic retrofitting be used to safeguard authenticity—meaning genuineness, not verisimilitude. Finally, the principles of reversibility and retreatability allow for the removal of interventions should they prove ineffectual, harmful, or inferior to methods that may be developed in the future.

The necessity of preserving architectural landmarks "in the richness of their authenticity . . . as living witnesses of their age-old traditions" was acknowledged officially when the International Conventions for Architectural Conservation (the Venice Charter) were adopted (Riegl 1964). One measure of authenticity is the amount of significant historic or cultural fabric retained by an object, be it a building, sculpture, fountain, or other resource. Thus, the primary conservation objective of maximizing authenticity is achieved through the preservation of historic fabric, a fundamental premise that should be observed when selecting seismic retrofitting measures for culturally significant structures. To the degree that portions of structures are missing or replaced, the authenticity of the whole is diminished, even though a replicated portion may preserve the integrity of the design.

Over time, buildings have suffered considerable losses of fabric from both natural causes, such as weather and natural disasters, and human causes, including warfare, vandalism, and inappropriate or clumsy restoration techniques. In the past, building elements often were replaced wholesale rather than repaired, and surfaces were renewed, not conserved, thereby reducing the authenticity of the whole. In particular, structural stabilization and seismic retrofitting of historic adobe buildings typically have involved the sacrifice of traditional, handcrafted structural systems and portions of adobe walls along with extensive use of structural materials that are mechanically incompatible with adobe. Concealment of retrofit measures (interventions) has been of paramount importance, and this principle has contributed to rejection of the time-honored, visible fixes traditionally used (buttresses, tie-rods, wall or joist anchors, cables, etc.). The priority placed on concealing structural interventions led to a disregard for the cultural values of the adobe walls themselves and their component parts. Exclusive concern with visual qualities has led to the preservation of architectural details at the expense of the whole in some instances—a phenomenon at variance with current conservation principles and practice.

A recent survey of adobe landmarks in California (Tolles et al. 1996) revealed that an additive or supplemental approach to structural stabilization had been used in only a few structures. The overwhelming majority of those buildings surveyed had been reworked with modern construction materials, neither supplementing nor replicating the early components in concept or detail. Little evidence of recording or salvage was found beyond the Historic American Building Survey (HABS) drawings completed as part of the Depression-era WPA program.

Retrofitting Objectives and Priorities

Adobe structure conservation efforts should take a focused, disciplined approach to the development of design options that are first consistent with safety and then with preservation of the historic fabric of the buildings. This entails a four-step process:

1. The structure is fully characterized;
2. important features and significant characteristics are identified;
3. an understanding of the structure in its historical context is established; and

4. design options that are creatively respectful of the structure's historic fabric are developed.

Once the prerequisite of life safety has been attended to, it is important to limit the extent of damage during seismic events. Efforts to improve the seismic performance of historic adobes are important not only *before* the next major earthquake but also afterward, as advocated by Feilden (1988), who stated that "we must always be aware that we live Between Two Earthquakes." The goals, after life safety has been assured, are to limit the damage to reparable levels during the most severe earthquakes and to limit the amount of cosmetic damage during moderate earthquakes. The order of these two objectives may be interchangeable. For example, it may be just as important to prevent cosmetic damage to surface finishes during frequently occurring moderate earthquakes as to ensure that a building remains reparable during infrequent major temblors. A variety of combinations of retrofit measures may be used to attain each of the three objectives.

In recent years, California, through its State Historical Building Code (SHBC) and Seismic Safety Commission (appendix C), has initiated a move away from the excessive intervention prompted by the Uniform Building Code (UBC 1997) for the seismic retrofit of adobes. The SHBC now provides other acceptable retrofit approaches for treating materials and structural systems that would have been considered either archaic or nonconforming under modern building regulations. We hope that other jurisdictions will follow suit and that a combination of scientific investigation and respect for conservation concerns will continue to expand the options available for the seismic retrofitting of adobes, options that take advantage of, rather than ignore, the existing material properties.

Chapter 2

Acquisition of Essential Information

The importance of adopting a holistic approach to building preservation cannot be overemphasized. As a practical matter, retrofitting strategies that are respectful of historic fabric cannot be devised without specific information on the fabric and other important architectural and historical features. It is necessary to identify precisely and in writing the features or qualities that should not be compromised. The need for comprehensive planning is clearly stated in a U.S. National Park Service publication dedicated to retrofitting historic sites for handicapped accessibility (Park et al. 1991:10): "The key to a successful project is determining early in the planning process which areas of the historic property can be altered and to what extent, without causing loss of significance or integrity. In order to do this, historic property owners and managers, working together with preservation professionals and accessibility specialists, need to identify accurately the property's character-defining features and the specific work needed to achieve accessibility. A team approach is thus essential."

Before the professionals who are charged with designing and implementing the seismic retrofit are authorized to proceed, the adobe building should be characterized fully and all the significant features and materials of the building identified. Recording and dissemination of the resulting report to all parties involved in a project should ensure that the essential information is shared. The optimal means of achieving this end is through preparation of a historic structure report.

The Historic Structure Report

In professionally managed projects, the historic structure report (HSR) is prepared by a multidisciplinary team. An invaluable resource, the HSR both provides information and guidance to those formulating interventions, including seismic strengthening, and analyzes "the emotional, cultural and use values in a historic building" (Feilden 1988:32). The HSR provides a comprehensive overview of the significance of the building and its components, as well as details about specific features and construction history. However, the practice of rating the significance of individual features can cause the sense of the whole to be lost in the recognition of the parts; therefore, the effects of multiple interventions must be assessed overall. As preservation engineer and architectural conservator Ivo Maroevic (1988:11) has observed, "It is dangerous to define the values of a monument, and thus also its identity, by evaluating single elements regardless of the unity formed by a building or settlement. It causes the separation of a part from the whole and the formation of a separate identity of the detail." The risk is that individual samples of significant historic fabric will be preserved at the expense of the design of the whole if an appropriate balance is not struck. This balance is not achieved by happenstance but through detailed knowledge and documentation of the building, its construction, and its alteration history.

Values Identification

It is critical to understand the various "values" of culturally significant or historic structures. The term *culturally significant*, as used in the Burra Charter on vernacular buildings preservation (Marquis-Kyle and Walker 1992), is increasingly preferred to *historic* because it is broader and clearly inclusive of both historical and architectural significance. For purposes of discussion, values are divided into those that reflect physical or visual qualities and those that do not. Architectural, aesthetic, or artistic values of structures fall into the first category, whereas spiritual, symbolic, associative, historical, or documentary values fall into the second. Archaeological values, including research potential, can fall into either category, depending on the nature of the archaeological resources.

Frequently, culturally significant structures possess values or derive significance from both of these arbitrarily defined categories. Mistakes or omissions in identifying the various values of a specific structure occur when undue weight is attached to one category of values or the other, reflecting the disciplinary background of the investigator. Evaluation by a multidisciplinary team avoids this problem. In practice, aesthetic (visually manifested) values are often emphasized over the less tangible ones (such as historical significance or research potential) because the public generally expects historical monuments to look attractive and conform to contemporary notions of good taste regardless of their historic appearance. This explains why so many historically important muraled surfaces of the California missions were painted out during the twentieth century, which, dominated by Bauhaus sensibilities, largely despised decoration. An example is the decoratively painted surface of Mission San Fernando Convento, which was painted over during the repairs that occurred following the 1971 Sylmar earthquake (figs. 2.1a, b). Some decorations have been "re-created" on the new surfaces, but the originals have been compromised.

Conservation theorist Cesare Brandi, discussing the aesthetic and historical duality of cultural property, observed that only the material, or what is necessary for physical manifestation, is restorable, not the historical aspects (Brandi 1977:2). Once lost, they cannot be recaptured through replication. If an adobe church that is considered significant because Father Junipero Serra said mass in it collapsed in an earthquake and was later reconstructed of new materials, it would not possess the same associative values as the original church. The archaeological site

Figure 2.1
Mission San Fernando, Mission Hills, Calif.: (a) painted surface before 1971 Sylmar earthquake (courtesy San Fernando Valley Historical Association), and (b) original painted surface covered by plaster and paint after earthquake repairs (photo courtesy David L. Felton).

(a)

(b)

might retain those values, however. Such is the case of the San Fernando Mission Church, which was demolished following the 1971 earthquake and reconstructed of new (non-adobe) materials. As is apparent from these examples, historical, documentary, associative, archaeological, and spiritual values are not always self-evident. Usually they cannot be clearly identified or recognized without research. Thus, an interdisciplinary approach is essential for the discovery and identification of such significant fabric and values.

Theorist Alois Riegl (1982) coined the term "age-value," an effect that is evidenced by the decay and disintegration of material or by the patina, to describe the visual quality conferred by age. The Bolcoff Adobe at Wilder Ranch State Park in California is an example of a building admired for its decrepitude (fig. 2.2). Age-value may or may not relate to significance and it may or may not be advisable to preserve it, depending on circumstances. For instance, the erosion of the base of an adobe wall, such as that exhibited by the Bolcoff Adobe, may threaten the stability of

Figure 2.2
Bolcoff Adobe, Santa Cruz, Calif.

the structure. Romantics who argue on aesthetic grounds for the retention of potentially hazardous, semiruinous conditions—such as retaining
and continuing to water vegetative wall coverings—are shortsighted, and
even irresponsible. Riegl said it well: "The cult of age-value contributes
to its own demise" (1982:33).

Historic Structure Report Formats

HSR formats are available from the sources listed in appendix D, all of
which recommend a multidisciplinary team approach. Preparation of a
comprehensive HSR is required by many public agencies and by some
important public and private grant-funding sources prior to interventions
involving major historic buildings. For buildings of minor significance,
a less comprehensive HSR may be sufficient. In instances where private
financing through local fund-raising efforts is relied upon and grant-writing
experience is lacking, financial contributors are often determined to see
so-called bricks-and-mortar results, not paper reports. Understanding that
reality, some sources of architectural conservation funding prefer to support preconstruction planning efforts, leaving support for the bricks-and-
mortar (construction) phase to local fund-raising efforts.

Minimum Information Requirements

This section outlines the minimum information needed to begin to plan the
design of a seismic retrofit for a historic adobe building in conformance
with architectural conservation principles and practice (see chap. 1). The
written report resulting from the accumulation of the following information represents an incremental or focused approach
to preparing a minimal historic structure report.

Historical significance statement

A historical significance narrative should articulate clearly the historical
significance of the building or complex of buildings, giving dates or
approximations of dates of construction. It should explain the social-
historical values represented to ensure that spaces and rooms (volumes
significant to social and/or architectural history) are not compromised by
intrusive measures, such as adding shear partition walls where no walls
existed previously. The narrative should assess the relevance of the
building and the people who built and occupied it in terms of themes,
such as Spanish colonization of the New World, the Mexican War of
Independence from Spain, subsequent Mexican colonial policy, Manifest
Destiny and the Mexican War, the California gold rush and statehood,
and for more recent times, the social trends and historical events that
have led to architectural revivals of archaic building materials.

History is not snobbish—important historical events have
taken place in humble kitchens, patios, bedrooms, *corredores*, and back
alleys, as well as in the architecturally embellished grand or public spaces
of churches, *salas*, lobbies, and ballrooms. There exists some confusion

about the difference between historical and architectural significance and architectural grandeur. A building can be significant, both architecturally and historically, without being grand or elaborately detailed, for example, the old Spanish Custom House in Monterey, California (fig. 2.3). The relative "humility" of adobe as a building material, as opposed to those materials traditionally termed "noble," has led to regrettable losses of historic fabric at the hands of those uninformed about history and vernacular architecture.

Types of questions to be answered through archival research include, but are not restricted to, the following:

- Who designed and built the building and in response to what needs?
- What role, if any, did its builders, occupants, or visitors play in historical events, for example, as participants in the Portola Expedition, the Anza Trek, the Hijar-Padres Colony, or in the establishment of missions, presidios, pueblos, or ranchos?
- How were these individuals representative of, or connected with, historical movements or themes, such as mission building, colonization, secularization, and rancho settlement?
- What historical events took place within or around the historic structure?
- How was life in and around the structure characteristic of the era?
- Why and how has the structure survived or been preserved as a witness to past events?
- What changes were made to the structure and in response to what historical events or human needs?

It is important to identify, if possible, the rooms or spaces in which historical events took place, as the preservation of their appearance at that time may be a priority due to historical association. For example, the public rooms on the ground floor of the Larkin House in

Figure 2.3
Custom House, Monterey, Calif. (courtesy of State Museum Research Center, California State Parks; photo by Robert Mortensen).

Figure 2.4
Larkin House, Monterey, Calif.

Monterey (fig. 2.4), used by Thomas O. Larkin, U.S. Consul to Mexican Alta California in the 1840s, are particularly significant in the context of the nineteenth-century American policy of Manifest Destiny promulgated by President Polk. Obviously, the level of effort expended should be commensurate with the importance of the building.

Historical-architectural significance statement

The historical-architectural significance narrative should describe the historical-architectural significance of complexes of buildings, individual buildings, and individual elements, features, spaces, and volumes of buildings. It should identify all elements of architectural distinction or importance worthy of preservation, such as those that define the character of the buildings, those representative of particular periods in the history of building technology, those that are unique or unusual or provide particularly early or late examples, and those that represent antiquated craft techniques.

The architectural significance statement should establish the construction sequence of building complexes and individual buildings, documenting the dates and construction techniques of original portions and subsequent additions and remodeling. This task entails both archival research and physical investigation to inventory and evaluate architectural features. Such research includes pictorial documentation (historical photographs, paintings, sketches, drawings, plans), written documentation (historical descriptive accounts, newspaper articles, specifications, building permits, contracts), and maintenance records, including accounts.

Inventory and evaluation of architectural features

The features, elements, materials, and spaces identified during formulation of the historical and architectural-historical significance statements should be physically inventoried, documented, and evaluated as to historical and architectural significance and integrity. This process involves physical investigation of the building to confirm or disprove the evidence of the historical and architectural record. The resulting inventory establishes what is and is not historic fabric worthy of preservation, documents the condition of the historic fabric, and makes recommendations for its conservation. It is imperative to note which elements or spaces are not significant or are of lesser significance, thus establishing a hierarchy, or order of priority to be consulted when devising the seismic retrofit scheme.

Concealed historic fabric

Identification of the visual, character-defining elements has been the subject of several monographs and at least one checklist (Nelson n.d.; Jandl 1988). This approach is problematic in that materials are identified as contributing to visual character but only their surface or superficial appearance is deemed important. Their authenticity as part of the documentary or historical value of the structure may be ignored. The difficulty of viewing a structural feature in an attic, for example, might cause it to be overlooked in the assessment procedure. Alternatively, the fact that a feature is structural and not decorative might cause it to be dismissed altogether. Brandi, dividing aesthetic values into structure and appearance, said that "the distinction between appearance and structure is very subtle, and it will not be always possible to maintain a rigid separation between the two" (Brandi 1977:2). He astutely observed that a change of structure could have an effect on appearance.

The significance of historic building fabric is not affected by its location or visibility. Lack of visibility of adobe bricks and mortar beneath a plaster surface does not diminish their value. Historic fabric in concealed locations beneath floors, in attics, and inside cavities can be very important for understanding the building's age, construction, and evolution and the building technology of the past. Observers have noted that in the recent past, serious mistakes have been made by preservationists' "replacing entire structural systems and concealed fabric, since they are not considered highly significant because of their lack of visibility" (Araoz and Schmuecker 1987:832). Another critic concerned about the adverse effects of seismic retrofitting on the cultural values of historic buildings notes, "In many cases, these values derive from the physical characteristics of the historic building including those of its structure. It is unfortunate to see that often these structural characteristics are considerably altered; hence, destroying part of the value of the building" (Alva 1989:108).

A positive example of the potential for preservation of original structural features is the retention of the original trusses at Mission Dolores, San Francisco, by architect Willis Polk following the 1906 earthquake (fig. 2.5). The similarity of this mission's Spanish Colonial roof trusses to those employed centuries earlier in the Andes is remarkable. It clearly demonstrates the continuity of Hispanic architectural traditions. Because so few original roof systems survive today, it is difficult to know with certainty what is or is not typical, atypical, or unique in Spanish and Mexican Colonial architecture in California.

Nonoriginal historic fabric
It is critical to identify and evaluate the physical evidence of cross-cultural adaptation of building technology added during subsequent periods. For

Figure 2.5
Mission Dolores, San Francisco, Calif., original roof support truss structure.

example, elements reflecting the adoption of Greek Revival features, such as the six-over-six windows at Mission San Juan Bautista, are typical of historic adobe buildings in California after about 1830. In New Mexico, so-called Territorial details became common after construction of the railroad. The preservation of obvious visual features is desired, along with the underlying structural elements, because all are authentic documents that provide information on the history of building technology.

Spaces and volumes

Not all architectural values are functions of structure or surface decoration. The organization of spaces—their shape, volume, linkages, and relationships—are as much a part of the architectural design and impact as surfaces. A recent monograph provides guidelines for ranking primary and secondary spaces based on visually derived data alone (Jandl 1988); however, the author recognizes the potential problem of a design professional's conducting the survey assessment alone, without benefit of the historical research required for preparation of a historic structure report. For example, the significance of a space presently used as a workshop, laundry room, or garage could easily be overlooked, and its identity as a rarely surviving attached or detached kitchen, such as the *cocina* of Rancho Camulos, could be missed completely.

Architectural surface finishes

The value of preserving early architectural surface finishes is more often recognized today than it was in the past. Technological advances have improved the feasibility of preservation, and the importance of original finishes has been demonstrated and appreciated. Architectural surface finishes other than mural painting—such as whitewash, tinted whitewash, paint, graining, glazing, scumbling, stenciling, marbling, lining, penciling, and the like—can be detected and conserved or replicated if necessary. Historic wallpaper can be conserved in situ or samples taken for replication, documentation, and preservation.

Finishes sometimes have the potential to be restored to the appearance they had at a particular, significant time, enhanced by the effect of a patina acquired over the years. Even when early surfaces beneath subsequently applied layers cannot or should not be exposed, they form a record of the modifications over time. These layers can be sampled to determine colors and treatments for the sake of scholarship or for possible replication on the existing surface. Through analysis of pigments and ground composition, paint conservators can formulate custom isolation barriers to protect early surface treatments while allowing for replication of such treatments on exposed surfaces.

Research at Missions San Juan Capistrano, San Juan Bautista, and other sites has demonstrated that humble spaces, such as *corredores* and workrooms, were sometimes embellished with decorative motifs by the mission neophytes (Neuerburg 1977). Some appear related to Native American rock art, while others reflect the meeting or blending of two cultures. Historical archaeologists working at the Mission Santa Cruz neophyte quarters uncovered original mud plaster that showed handprints and graffiti made by the virtually extinct local Ohlone Indians (the Aulinta and other local tribelets) beneath layers of later mud plaster and whitewash. A conservator experienced in the technique professionally conserved these marks through reattachment of the plaster to the mudbrick surface.

It is not unusual for surface finishes to provide insight regarding inhabitants of historic buildings. For example, the exterior mud rendering on the rear wall of the Boronda Adobe in Salinas, California, is decorated with fanciful graffiti scratched into the surface. A drawing of a large, grinning face with sombrero and mustache that is signed by one of Boronda's sons can be distinguished in raking light. This humorous personal expression from the past enlivens the bricks and mortar with humanity; Watkins noted it in his report almost thirty years ago and recommended its continued preservation (Watkins 1973:4). At the De la Guerra Adobe, in Santa Barbara, California, historical archaeologists have also inventoried and documented early graffiti encountered in the investigation process (Imwalle 1992).

In addition to decorative painting on plaster, wood, wrought iron, or stone, architectural features may have been gilded, carved, or otherwise embellished. Such features as corbels, arches, altar railings, ceiling *vigas*, altarpieces, and pulpits (fig. 2.6) may require attention to determine whether they need to be removed during retrofit procedures or can safely remain in place with adequate covering.

Not all surface finishes are created equal. Many times the exterior rendering of historic adobe buildings represents a functional, sacrificial protective coating that has been added to or replaced many times over the years. If the surface material is original, very early, or distinguished by unusual workmanship or materials, it may be desirable to retain as much of the finish as possible. However, the value of preserving a surface finish needs to be weighed carefully against other competing factors.

Muraled surfaces

In the process of investigating and identifying the building's historic fabric, murals of considerable art-historical importance may be encountered, either on the surface or concealed beneath later layers or additions. They may be deteriorated or adversely affected by their present conditions. If there is any question about their condition, extent, or potential for future adverse effects from seismic retrofitting procedures, a conservator with expertise on murals should be consulted for advice on protective measures.

In the Spanish and Mexican eras in California, prior to 1850, mission churches,

Figure 2.6
Pulpit at Mission San Miguel, San Miguel, Calif.

conventos, rancho headquarters, and residences were decorated with wall paintings that are now quite rare (Neuerburg 1987). While few of these are readily visible today, there are indications that figurative painted decoration may be preserved beneath later layers of paint or plaster at many sites. Until the 1970s, the *convento* at Mission San Fernando was distinguished by the presence of extraordinary murals painted by mission neophytes that reflected both Native American and European art traditions. During the Great Depression, the murals were recorded by artists from the Index of American Design (fig. 2.7). However, the paintings are no longer visible; they were plastered and painted over as part of the renovation and seismic retrofitting of the *convento* after the 1971 Sylmar earthquake. Some of the murals have been reinterpreted on the new surfaces (fig. 2.8). It is possible that some of the original paintings lie beneath the plaster, awaiting future conservation. But their potential can only be realized if their existence is known and their significance acknowledged and respected. When such decoration is not visible, historical-architectural research might produce descriptions of similar paintings or early photographs might be located that would indicate their existence beneath later finishes.

Historical-archaeological resource evaluation

The role of historians and architectural historians has been discussed relative to the HSR and the portions that are essential for the architect's and engineer's information. The role of historical archaeology is important because archaeological resources may be adversely affected by seismic retrofit procedures or testing that involves subsurface exploration of foundations or the geology of the site.

Figure 2.7
Mission San Fernando, Mission Hills, Calif.: original murals as recorded in the Index of American Design (courtesy National Gallery of Art).

Figure 2.8
Mission San Fernando, Mission Hills, Calif.: murals reinterpreted following 1971 earthquake repairs.

Provisions of the California Environmental Quality Act protect significant archaeological resources, and sections of federal law protect certain Native American areas, particularly burial sites. At missions and other churches, burials may be encountered beyond the walls of the cemetery. For example, more than one cemetery existed at sites such as the missions at San Diego and San Antonio, and burials can be found beyond current cemetery walls at Missions Santa Cruz and San Juan Bautista. Often, the location of the cemetery is as yet unknown, as is the case at Mission La Purisima Concepcion in Lompoc, California.

Besides the possibility of burials, historic adobe sites are usually archaeologically sensitive and, in the case of seismic retrofitting measures that involve foundation work, generally require protection from ground-disturbing activities. Borings for geotechnical studies fall into this category, as well as excavations for footing inspections.

Former mission neophytes provided the bulk of the labor required to build the historic adobe buildings of California both during the mission era and afterward, when they found employment on the ranchos and in the pueblos, often living as servants on the premises with the Californios. Thus, most historic adobe building sites potentially require Native American–related archaeological sensitivity and investigation. Depending on the sensitivity of the potential resources, it may be advisable to have local Native American representatives on site as monitors to prevent or resolve misunderstandings about the protection and disposition of excavated materials. Failure to conform to laws that protect archaeological resources on either public or private property can lead to unforeseen and often unpleasant consequences.

Virtually all historic adobe sites have culturally significant archaeological deposits that could further the knowledge of life in the past and thus require professional evaluation to determine their integrity and extent. Excavation of the earth at these sites, both within the structure (because many historical adobe structures had earthen floors originally) and outside the buildings, necessitates the early participation of a historical archaeologist. Tasks to be performed include evaluation of the research potential of the site, identification and protection of any significant historical resources, and, if necessary, design of a mitigation program to offset any unavoidable negative effects on archaeological resources. The historical archaeologist will need to refer to the historical and architectural research performed. If significant historical resources are detected initially through survey, testing, or archival data, the design professionals can try to work around them.

Inventory team leadership and composition

Architectural historians, historical archaeologists, and preservation or historical architects contribute to the inventory process under the direction of a leader who represents one of the disciplines and possesses specialized knowledge of the architecture of the period. As historical archaeologists are adept at interpreting the sequence of events and establishing which features are contemporaneous—a skill derived from extensive training and experience dealing with stratigraphy—specialists in this field have often emerged as team leaders in building investigations here and abroad. France has been an innovator in the field, and UNESCO cultural heritage officials have developed an international training program. California State Parks has assembled multidisciplinary teams led by a historical archaeologist for conducting investigations of historic buildings. At times, architectural historians or historical architects assume the leadership role.

Regardless of the discipline represented, the specialist responsible for the inventory should be experienced with the identification of building techniques ranging from the archaic to contemporary and should be familiar with construction methods, materials, and tool marks and the means used in their replication. Many historic adobe buildings have been "restored" to one degree or another though the use of replicates or wholly reconstructed elements. These require accurate identification because they may be the work of craftspeople or groups whose work has acquired significance over time, such as the Civilian Conservation Corps, which reconstructed Mission La Purisima Concepcion in the 1930s, and the Mexican friars who rebuilt Mission San Luis Rey in the 1890s. These artisans merit consideration in their own right. Other replicated features may possess no real significance and may be candidates for replacement by an anchor bolt, cable, or other retrofit device. It is crucial that the design team be provided with accurate information about what is *not* important, which will allow them the maximum possible leeway in formulating a retrofit program.

The level of investigation varies with the size and importance of the building, the budget, and the professional assistance available, but the type of information needed to make informed decisions does not. It may seem less expensive and more expedient to instruct a designer that "everything" about a historic building is historically or architecturally significant without going to the trouble and expense of identifying the truly significant features. This approach can render a designer's task nearly impossible to perform responsibly, and as a result the project becomes unnecessarily expensive. Alternatively, failure to identify the qualities that determine a building's historic value can result in the loss of historic significance and designation if the designer unwittingly compromises those qualities.

Summary

The preceding issues need to be considered and the information assembled and synthesized for the design team before proceeding from preliminary concept drawings or design development to final working drawings, construction documents, permits, and environmental or historic resources commission reviews. Each historic adobe building is different and has individual features that should be dealt with on a case-by-case basis.

The Historic American Buildings Survey documented some historic adobe buildings in the 1930s and some documents were prepared by architects before the start of a major restoration effort. In such cases, the existing data may need only to be located, confirmed, and updated. It is worthwhile to conduct the research necessary to find previous plans, drawings, and photodocumentation to avoid wholly "reinventing the wheel." It is much more efficient and less expensive to confirm measurements on an existing plan than to prepare existing-condition drawings anew. In addition, good maintenance and rehabilitation records exist for some sites, and plans, specifications, and contracts can be located easily. Local building department files should be searched for such records.

While it is beyond the scope of these guidelines to deal with interventions other than seismic retrofitting, preparation of a historic structure report is indicated when large-scale repairs or modifications are necessary to stabilize a structure. A historic structure report is advisable when a major change of use, termed an *adaptive reuse*, is proposed that is likely to alter the physical manifestations or the record of the flow of history of the building. When major modifications are proposed, even for the sake of restoring the building to an earlier and possibly more structurally viable condition (such as reconstructing missing adobe transverse walls), the process of reversing the changes sets back the historical clock, so to speak. Then this interference, however beneficial the end result is intended to be, alters the artifact forever. Permanent alterations necessitate a thorough documentation of existing conditions before the proposed changes are effected. These types of changes are carefully scrutinized by governmental preservation officials and require well-documented justifications.

Chapter 3

Practical Application: Retrofit Planning and Funding

When owners and managers of historic adobe buildings postpone retrofit planning until an emergency situation arises, they risk compromising irreplaceable historic fabric. Precipitous decisions that are not necessarily cost effective can result from haste and the lack of necessary information.

Engineers experienced in the field of historic structure preservation treat all of a historic building's fabric as significant when specific information to the contrary is lacking. As an engineer observed in a report evaluating the structure of Colton Hall (the site of California's constitutional convention) for the City of Monterey, "Most of the building's fabric must be assumed to be primary until proven otherwise" (Green 1990:16). In the absence of the historical and architectural information necessary to identify significant historic fabric, the designer's task is complicated by necessary, if excessive, caution. This makes the designer's task more costly and time consuming.

By following systematic advance planning procedures unnecessary work can be avoided, and the project can be implemented more rapidly. For example, the historic structure report prepared for Colton Hall aided the engineer in avoiding irreversible impacts to significant historic fabric. Similarly, intensive architectural investigations begun after completion of the seismic retrofit of the Casa de la Guerra in Santa Barbara, California, revealed opportunities for different detailing than was possible beforehand. This illustrates the value of intensive physical investigation before the initiation of retrofit design activities (Imwalle and Donaldson 1992).

Secondary, windfall benefits also sometimes accrue. An unforeseen but important benefit to the San Juan Capistrano Mission Museum came out of the advance planning and research performed preparatory to the mission's seismic retrofit program (Magalousis 1994). The historical, archaeological, and architectural investigations necessary to collect the information required by the design team generated important new information about the mission. This in turn resulted in revision of the museum's interpretive program and its presentation of the architectural history of the mission. The information-gathering and synthesis process was a revitalizing force at the mission and enhanced the visitor's experience at this major tourist destination.

Preliminary Condition / Structural Assessments

Before making decisions regarding project planning or personnel, the owner or site manager should take the essential preliminary step of defining the scope of the proposed project. It is advisable to retain the services of a preservation or historical architect and an engineer who is experienced with adobe buildings to prepare written condition and structural assessments and preliminary cost estimates. Such reports provide information in general terms on the level of intervention necessary to secure the building. No commitment to retaining the ongoing services of the professionals consulted need be made, but their findings should influence decisions, depending on the magnitude of the problems encountered. Additional opinions or a peer review by another qualified professional architect or engineer can be sought, if necessary (see "The Seismic Retrofit Planning Team," later in this chapter, for details on the personnel who should be involved in the planning stages).

The preliminary condition and structural assessments should provide the information necessary to determine whether the building is in such condition that the design of a seismic retrofit project can proceed directly or whether other preservation treatments, such as additional structural stabilization or repairs, are needed first. The building must be physically surveyed, findings made, and recommendations formulated from the perspective of both preservation architect and engineer. The usual training received by engineers is focused on structural and life-safety concerns, whereas the historical architect is trained to understand what needs to be done—and by whom—to safeguard the historical, architectural, and archaeological features of the structure. This architect is also better able to deal with the important aesthetic and design issues.

With condition assessments of the adobe building and cost estimates in hand, the owner, manager, committee, or board responsible for the care of the adobe structure will be in a better position to begin planning the seismic retrofit or other type of preservation treatment.

Choosing the Appropriate Preservation Treatment

Seismic retrofitting is one specific type of structural stabilization or intervention that is considered part of a preservation treatment. Terms for alternative interventions, such as *rehabilitation, restoration, reconstruction,* and *preservation,* are descriptive of various approaches to the conservation of a historic structure, and are defined in appendix F. Whatever approach is adopted, all treatments should embrace the conservation goal of maximum retention of historic fabric to preserve authenticity and should be preceded by a systematic, multidisciplinary investigation of the building. Seismic retrofitting measures can be included as part of any of these broad-scope preservation programs or stand alone as specialized stabilization. A preservation or historical architect, in consultation with the owner or site manager, can provide guidance in selecting the appropriate treatment. In some cases, the condition and structural assessments completed prior to commencing a seismic retrofit project may conclude that more than a minimum program of seismic retro-fitting is indicated to rectify hazardous conditions or accommodate proposed new uses for spaces. Temporary operations, such as shoring and other short-term seismic strengthening measures, may be recommended to stabilize the building while long-term planning and fund-raising for the project are accomplished (Harthorn 1998).

Special Circumstances

Critical conditions

Cyclical maintenance has long been acknowledged as being of critical importance in the ongoing preservation of buildings. Regular mainte-nance is important for adobe buildings, particularly those located in regions of high seismicity. The findings of international earthquake conferences on earthen buildings in seismic areas were that well-maintained adobe buildings have a greater chance of survival than those in poor condition.

Serious building conditions that require attention when considering a retrofit program include

- basal erosion (coving at the wall base);
- poor site drainage;
- excessive moisture in walls, especially those covered by hard, impervious cement-type renderings;
- additional wall penetrations at structurally critical areas, such as corners and between original openings;
- missing interior transverse walls;
- large areas of in-fill composed of incompatible materials of differing physical properties;
- absence of, or poorly attached, roofing;
- absence of connections that provide continuity between adjacent building elements; and
- evidence of previous severe earthquake damage that was only cosmetically repaired.

If the preliminary condition assessment and structural analysis identifies such conditions as critical, the need for a more far-reaching program to rectify deficiencies may be indicated. However, regardless of the level of intervention indicated or the program adopted, safety and historic fabric retention remain high-priority goals.

Limited destructive investigation

The need for comprehensive information about a building's structural condition is a good reason to undertake the extensive physical investiga-tion of a building required for a historic structure report (see chap. 2). Evidence of previous earthquake damage or other conditions, such as moisture damage at wall bases that may affect seismic performance, is often camouflaged by cosmetic repairs. The architect or engineer who performs the initial evaluation of the building may recommend limited removal of wall renderings in order to understand the past seismic behavior of the building and to investigate possible moisture damage. It is important to conduct such efforts because, without a thorough investigation, it is possible to reach erroneous conclusions about a building's prior performance in earthquakes and its existing condition. However, it is important also to insist that such investigations be limited and the findings be thoroughly documented because of the possibility of serious damage to historic fabric and surface finishes.

Testing

An architect, engineer, or architectural conservator may recommend various types of tests to obtain accurate information about important materials concerns. Included in this category are tests to analyze mortar, adobe, or fired bricks to determine their composition and material properties (strength, modulus of rupture) and geotechnical testing to identify soil conditions and verify geological formations and hydrological conditions below grade. The latter are indicated if any evidence of settlement or foundation deficiencies is observed.

Use issues and retrofit concealment

The existing or proposed new use of a building is an important consideration in designing a seismic retrofit. Not only are engineering considerations important—such as expected dead and live loads—but also the use to which a building will be put influences the degree to which retrofit devices must be concealed. In a house museum, for example, it is important to maintain the surface appearance of the building for reasons of interpretation. In such cases it is crucial to differentiate between that which is real and authentic, and should be preserved intact, and that for which only the *appearance* of age is important. In the latter case, historic fabric details can be replicated, if desirable.

Present and potential uses should be considered in a conservation approach to retrofitting, and the extent of concealment of retrofit devices can change with a change of use. For instance, at Colton Hall, city offices currently occupy the first floor, which has been slated for conversion to a schoolhouse museum. Levels of visibility acceptable in an office environment, where steel reinforcement of joists go virtually unnoticed, may be unacceptable in a museum. In recently completed work, seismic retrofits were boxed in and concealed. A retrofit designed for current use should anticipate possible future changes in the use of the building and be removable to facilitate redesign at some later time. It is important to take a long-range view of the structure and understand that its use may change several times in the future, just as it may have changed in the past.

Generally speaking, the greater the amount of concealment required by the proposed retrofit, the greater the potential intrusion upon or loss of historic fabric. When invisibility of the seismic retrofit is demanded, the tendency has been to remove historic fabric—either by channeling into or replacing it—and to wholly conceal the new element within the historic walls, instead of adding or attaching the new elements to the building. The impact of devices, elements, or features incorporated into the fabric of historic structures is usually far greater than that of those added or attached to historic buildings. An example is concrete bond beams. Their installation usually involves removing the existing roof, including its surfacing, sheathing, and framing. Courses of adobe brick are also either removed entirely or material is removed from the upper courses to form channels in which the steel-reinforced, poured-in-place concrete bond beams are installed. In the course of removal, hand-hewn timbers can split, and if termite- or dry rot–damaged wood is discovered, it is discarded, rather than repaired. Thus, invariably, historic elements of the roofing system are lost.

When artistic and historic murals or other wall paintings are encountered, the need for undisturbed preservation may require the use of specially concealed retrofit measures. A delicate balance has to be achieved between satisfying the need to preserve the comparatively plentiful historic fabric of the adobe walls versus the rare muraled surfaces such as those of Missions San Miguel and Santa Ines (figs. 3.1, 3.2). As others have said, "The problem for the seismic retrofit of historic structures is to find the balance of interventions that reduces the risk for injury or property damage to an acceptable level without unduly destroying the historic fabric" (Thomasen and Searls 1991).

Retrofit Opportunities

Earthquake damage repair

A retrofit opportunity presents itself when earthquake damage must be repaired before a building is declared safe for occupancy. Damage caused by the 1989 Loma Prieta earthquake to the Boronda Adobe, in Salinas, prompted the Monterey County Historical Association to engage an engineering firm and a preservation architect to recommend repair measures and seismic strengthening. The building had been rehabilitated fairly recently, was

Figure 3.1
Interior murals in Mission San Miguel, San Miguel, Calif. (courtesy Mission San Miguel).

Figure 3.2
Interior murals in Mission Santa Ines, Solvang, Calif.

well maintained, and its historical features had been evaluated and documented prior to rehabilitation. The Federal Emergency Management Agency (FEMA) contributed to the repairs and retrofit measures. Similarly, damage from the Northridge earthquake of 1994 resulted in the seismic retrofitting of the De la Ossa and Andres Pico adobes. The Del Valle Adobe at Rancho Camulos, in Piru, and the Leonis Adobe, in Calabasas, have already been retrofitted, the latter immediately following the Northridge earthquake (fig. 3.3).

Reroofing

Seismic retrofitting often involves the modification of the roofs and attic spaces of buildings, activities that can increase installation expense if access to these areas is difficult. However, it is imperative to maintain sound roofs on adobe buildings to prevent damage to the mud brick walls by water leakage. If there are any indications that an adobe building needs reroofing, incorporation of seismic retrofitting into reroofing plans should be considered seriously.

In 1991–92, the roofs of several historic adobe buildings in Monterey and San Benito counties, including Mission San Juan Bautista and Casa Abrego, were replaced (Craigo 1992). Although the owners were officially advised of the advantages of seismic retrofitting when replacing roofs, they declined to do so, and an economic opportunity was lost. The owners of the First Federal Court Adobe in the Monterey Old Town National Historic District, however, recognized the economic benefit of retrofitting while reroofing and engaged an engineer experienced with adobe buildings to design and oversee the retrofitting of the building (Monterey Historic Preservation Commission 1992).

Americans with Disabilities Act compliance measures

In the United States, historic buildings open to the public are required to comply with the provisions of the federal Americans with Disabilities Act (ADA) regarding public accessibility. In some instances, substantial modifications to historic buildings and sites are necessary to ensure access. The modifications must be carefully planned and budgeted and are often reviewed by local historical resources commissions. Before initiating the design work required for ADA compliance, it might be economically

Figure 3.3
Leonis Adobe, Calabasas, Calif. (courtesy Tony Crosby).

advantageous to consider inclusion of a seismic retrofit program. Some California communities, such as Sonoma, have linked seismic upgrading with ADA compliance. Local jurisdictions can form assessment districts to provide funding to assist owners in financing compliance measures, which is usually in the form of low-interest loans or sometimes outright grants (see "Securing Funding," later in this chapter).

The Seismic Retrofit Planning Team

Planning for the seismic retrofit of a historic adobe building requires the joint participation of a multidisciplinary group of technical specialists. The preservation architect and the engineer form the nucleus of the design team. A preservation professional or conservator can be added to the core team if the architect chosen is not a specialist in historic preservation of adobe buildings. All members of the core team require direct contact with the client as well as with each other.

Preservation architect

Preservation, or historical, architects have specialized historic preservation or architectural conservation experience and are professionally trained to coordinate large-scale historic preservation projects. They understand the need for and benefit of working with a multidisciplinary team to address the peculiarities of historic sites, such as the presence of archaeological deposits. A preservation architect familiar with the properties of archaic building materials and systems, such as unreinforced masonry made of adobe, stone, and *ladrillo* (the materials of historic adobe construction in California and the Spanish Colonial Americas), is preferable to an architect without conservation expertise.

Most professional architects are generalists who are broadly trained to plan, organize, and coordinate major projects—and when specific needs are recognized, to consult specialists for more detailed information. Therefore, should the services of a specialized historical architect prove to be unavailable or undesirable due to distance, budget, or other factors, a conventionally trained architect could be retained who would bring in specialists (architectural conservators, materials scientists, historic preservation consultants) to help identify and deal with unusual constraints. Such an architect may require professional assistance when dealing with preservation issues involving California's State Historical Building Code, or the local equivalent, and the standards and guidelines followed by historic preservation review boards and funding agencies.

Why is it necessary to engage an architect, with or without historic preservation training, when seismic retrofitting is an engineering problem? Architects are trained not only to oversee project planning but also to anticipate the possible consequences of altering existing buildings, especially with regard to the visual or aesthetic qualities. A preservation architect brings the additional qualifications of understanding the value of preserving historic fabric, knowing how to go about it, and being familiar with the appropriate specialists to entrust with various tasks. In contrast, engineers are adept at solving structural problems efficiently and cost effectively. However, all of the parameters and constraints must be clearly stated in order for them to perform their task with the exactitude characteristic of those in this profession. Generally speaking, some architectural conservation needs typical of adobe buildings, such as elimination of water intrusion, are not usually part of an engineer's expertise. The architect assesses all the variables and sets out the constraints or parameters within which the engineer will work.

Some may question the need to engage an architect to deal with a "simple" mud-brick building. Of the reasons professional architectural services are recommended, the most compelling relates to the historical and architectural importance and rarity of Hispanic-era architectural heritage, particularly in California. Few historic adobe structures remain, and the authenticity of those has been diminished over the years through insensitive rehabilitation, earthquake loss, and over-restoration. In fact, some scholars who have surveyed Spanish Colonial architectural heritage in the New World have dismissed California's architectural contributions as being unworthy of consideration due to their perceived lack of integrity and losses of historic fabric (Thomas 1991:119–149; Maish 1992:29). What historic architecture does remain requires professional treatment to assure preservation for future generations.

Sources of information in California regarding preservation architects experienced with historic adobe buildings include the following (see appendix E for further details):

- California Office of Historic Preservation
- National Trust for Historic Preservation
- Heritage Preservation Services, National Park Service
- American Institute of Architects

When engaging a preservation architect, it is advisable to contact agencies responsible for historic adobe sites for references, including the California Office of Historic Preservation and the Heritage Preservation Services division of the National Park Service. Both agencies are stewards for large numbers of historic adobe buildings.

Regardless of whether an architect undertakes project planning and supervision, or an owner, manager, board, or building committee representative assumes this responsibility, certain issues will require careful consideration, depending on the nature of the site and the scope of the project. Primary among these are concerns about identification and conservation of historic fabric, identification and preservation of archaeological deposits, and the advisability and feasibility of preparing a historic structure report (see chaps. 1 and 2).

Engineer

A structural engineer who specializes in historic buildings, or if possible in earthquake engineering or seismic retrofitting of adobe buildings, should be selected as one of the principal members of the planning team.

The California State Office of Historic Preservation (OHP) and the Heritage Preservation Services division of the National Park Service, which have sponsored two conferences on seismic retrofitting of historic buildings with the Western Chapter of the Association for Preservation Technology, may be contacted for the names of engineering specialists who have experience in this area.

Social historian

Historians, or certain historical archaeologists possessing a firm foundation in historiography, familiarity with local and regional archives, and experience with the Spanish Colonial and Mexican Republic eras, can compile the necessary data and formulate a meaningful historical evaluation of the adobe structure under consideration. Individuals with the necessary expertise can be located in a number of ways (also see appendix E):

- The OHP provides a list of historical resources consultants, which is available from the California Historical Resources Information System (CHRIS).
- The California Council for the Promotion of History publishes the Register of Professional Historians, which lists historians specializing in early California history.
- The California Mission Studies Association publishes a directory listing specialists in the history of the period.
- In the Southwest, the Southwestern Mission Research Center in Tumacacori, Arizona, and other agencies operating adobe sites may also be contacted for the names of local specialists.
- University faculty members can often supply names of qualified social historians who specialize in the Spanish and Mexican eras of early western history.

Architectural historian

Architectural historians who are familiar with the architecture of this era are ideally qualified to research and evaluate the architectural-historical significance of historic adobe buildings. The Office of Historic Preservation's referral list includes qualified architectural historians, and the Society of Architectural Historians may be contacted for member specialists. University and college faculty (active or retired) who specialize in Latin American architectural history may also be consulted. The California Mission Studies Association publishes a membership directory that lists practitioners in the field. The Southwestern Mission Research Center in Tumacacori, Arizona, is another source of information (see appendix E for more information).

Conservator

Some wall painting and architectural surface-finish conservators are experienced in the conservation of murals on adobe, on mud-rendered

surfaces, or on the more conventional lime plasters. The American
Institute for Conservation of Historic and Artistic Works, the Getty
Conservation Institute, or the Conservation Center, National Park
Service, Santa Fe, New Mexico, may be contacted for information or
references. Because alterations necessary to retrofit structures may
loosen significant renderings or finishes and threaten their adhesion,
a conservator should be consulted

- to confirm whether or not significant surfaces are present,
  and if so,
- to determine their extent and condition, and
- to recommend treatments that will stabilize the surfaces to
  withstand the installation of seismic retrofitting measures.

Architectural conservators trained or experienced with earthen
materials construction may be located through the sources mentioned here
and through the earthen architectural training programs organized by
CRATerre-EAG and ICCROM (see appendix E).

Historical archaeologist

The Register of Professional Archaeologists (RPA) is a directory of pro-
fessionally qualified archaeologists, some of whom have considerable
experience with historic adobe structures. The Society for Historical
Archaeology, the Society for California Archaeology, the California
Mission Studies Association directory, and the Southwestern Mission
Research Center may also be consulted for information about specialists.
Some historical archaeologists specialize in building investigation.
Contact the California Historical Resources Information System (CHRIS)
or the Santa Barbara Trust for Historic Preservation for information.

Securing Funding

Funding problems are one of the greatest deterrents to seismic retrofit
projects everywhere. Commercial property owners cannot raise rents
enough to cover the cost of a largely invisible upgrade. Similarly, a historic site cannot raise visitor fees sufficiently to finance a seismic retrofit
program. Historic adobe buildings in California may possess an advantage in competing for grant funding because of their relative scarcity,
their importance as relics of the state's earliest settlement, and their
educational and tourism potential. Unlike the preponderance of late-
nineteenth-century, urban, unreinforced-brick commercial buildings, historic adobe buildings were adapted to their sites and the changing size of
families over time, making no two alike. Each is a unique architectural
expression of the past and a surviving symbol of cultural change.

Historically, some nonprofit organizations, such as churches,
have relied on securing professional services gratis or at reduced rates
from building committee or board members, parishioners, historical
societies or preservation organizations, archaeological societies, college

students, and volunteers. In some instances, avocational archaeologists or student interns may be recruited to perform some tasks under professional direction. However, certain personnel substitutions or supposed economies, such as consulting a general contractor in lieu of an engineer to assess structural vulnerability and prepare retrofit designs for unreinforced historic adobe buildings, are not advisable and can raise liability issues in the event of casualties resulting from an earthquake (see "The Seismic Retrofit Planning Team," earlier in this chapter).

Government funding sources

Potential sources of project funding include historic preservation grants from California State Parks bond funds that have been approved by the electorate. Private, nonprofit agencies, as well as governmental agencies, may apply for such funds through the Office of Historic Preservation's California Heritage Fund. Seismic retrofit of unreinforced masonry buildings remains a high priority for receipt of federal historic preservation grant funds in California. Other smaller grants may be applied for through certified local governments and are awarded by the OHP. The Western Regional office of the National Trust for Historic Preservation makes small planning grants in California from a California-specific fund.

Federal Emergency Management Agency (FEMA) Hazard Mitigation Grant Programs provide matching grants to private owners for retrofitting in certain disaster-stricken regions. FEMA also provides disaster relief to public and private, nonprofit, organization-owned buildings, including funding for seismic retrofit procedures. In those communities, federal Community Development Block Grant and Economic Development Assistance funds may be available, if the building is a tourist attraction or is considered blighted.

Some local governments administer hazard mitigation loan programs with funds derived from establishment of assessment districts. The Small Business Administration (SBA), the federal agency that provides loans to private building owners following a federally declared disaster, will increase loans up to 20% for hazard mitigation expenses, including seismic retrofitting of earthquake-damaged buildings. Use of federal or state funding necessitates compliance with the "Secretary of the Interior's Standards for the Treatment of Historic Properties" (appendix F).

Tax credits

For commercial, income-producing properties—such as the Leese-Fitch or Salvador Vallejo adobes in Sonoma, California—that are on the National Register of Historic Places, the Federal Historic Preservation Tax Incentives program's 20% investment tax credit can be useful. Another way in which owners of historic buildings in California can reduce their property tax liability is by application of the Mills Act. This act allows a city to enter into a contract with the owner to reduce taxes by changing the way the tax assessor calculates the property tax. In return, the owner agrees to protect and preserve the historic property. One way this can be done is by installing seismic retrofitting.

Private funding sources

Private and local community foundations have made grant awards for seismic retrofitting and earthquake repair. Following the 1989 Loma Prieta earthquake, the Community Foundation of Monterey County funded earthquake repairs at the Casa de Soto from the Doud Fund. The Skaggs Foundation assisted Mission Dolores with retrofit costs after the Loma Prieta earthquake, and Mission San Fernando with repairs and retrofitting after the Northridge earthquake of 1994. The Kresge Foundation supports "bricks and mortar" projects on a challenge grant basis. Mission San Gabriel (figs. 3.4a, b) was braced and shored up following the Whittier Narrows earthquake of 1987, using funds provided by the Grant Program of the J. Paul Getty Trust, which accepts grant applications for planning and implementing architectural conservation of buildings that have been designated as National Historic Landmarks (National Landmark status is distinct from listing in the National Register of Historic Places).

The importance of advance planning

Fund-raising and grant-writing experience indicates that grant funding available from architectural conservation and historic preservation sources is increasingly contingent upon evidence of advance planning procedures. Attempts to cut costs by omitting steps necessary to preserve authentic historic fabric may lead to the rejection of the grant applications and the resultant loss of the time and money expended in the application effort. Grant funding is available to underwrite the preparation of planning documents including preservation plans, adaptive-reuse feasibility studies, condition and structural assessments, conservation studies, historic structure reports, and master site plans. Well-planned, thoughtful, and reasonably phased projects have been shown to offer the best chances of success both in economic and cultural preservation terms.

Figure 3.4
Mission San Gabriel, San Gabriel, Calif.: (a) bell tower before the 1987 earthquake (photo courtesy David L. Felton), and (b) bracing and shoring of the tower following the 1987 earthquake.

(a)   (b)

Chapter 4
Overview of Engineering Design

Many adobe buildings have survived major earthquakes while sustaining only minor damage. Others have suffered a considerable amount of structural damage. Some earthquake-damaged buildings have been repaired over the years, but many damaged structures were simply abandoned. The nature of adobe as a building material and the geometric configuration of buildings in which it is used make adobe structures unique building types. Adobe buildings differ from buildings made of other materials; therefore, the nature of adobe and how it is used as a construction material must be considered in the design of seismic retrofit strategies.

Adobe masonry walls are built up using unfired earthen bricks that are set in a mud mortar. The soil used for the bricks typically has a clay content that ranges from 10% to 30%, and organic material such as straw or manure is usually added before the bricks are formed. The organic material helps to reduce shrinkage and to minimize the formation of shrinkage cracks that usually occur as the bricks dry. The mortar is usually composed of the same soil material as the bricks, but it may not contain similar organic materials. Mortar is almost always weaker than bricks because rapid drying during building erection can lead to shrinkage and cracking of the mortar.

It is normal for adobe building walls to be cracked as a result of this shrinkage, inadequate foundations, and/or differential settlement. Part of the cultural tradition in the areas where adobe buildings are the vernacular architecture is to repair cracks periodically when renewing the mud-plaster surface finishes. Over the years, adobe structures may have undergone major additions and modifications and the building configuration may have been changed considerably. During moderate to major earthquake ground motions, most adobe buildings have experienced additional cracking, and the repair of such structures was an integral part of local tradition and culture.

Historic adobe buildings were usually built with thick walls but had a roof system that was poorly attached to the walls. The thick walls of historic adobe buildings are important features that enhance seismic stability; however, the roof should always be properly attached to the walls. From an engineering perspective, the characteristically unique stability of adobe buildings can only be fully realized if the walls are permitted to crack during movements caused by an earthquake. In fact, it is virtually impossible to prevent adobe buildings from cracking under these circumstances. It is therefore imperative that the theoretical basis for an engineering analysis include consideration of the dynamic performance of cracked adobe structures.

Principles of Seismic Design

The comprehensive engineering understanding of the seismic performance of structures is of recent vintage. It is only in the past century that an understanding of how structures respond in earthquakes has begun to emerge. Historical building practices evolved through the accumulation of experience gained by trial and error. The first measurements of ground motions during damaging earthquakes were not made until 1933, whereas it was only in the 1970s that the first recordings were made of a building responding to an earthquake that caused damage to that building. The first engineering procedures for seismic design were not formulated until early in the twentieth century, although some sporadic attempts were made previously. These efforts were then augmented by an accumulation of construction details that were asserted to give satisfactory seismic performance.

Following the emergence of modern construction methods, in which steel and reinforced concrete replaced brick and stone as principal building materials, structural designs were developed that would allow buildings to withstand severe environmental loads (wind and earthquake) and perform in predictable and acceptable ways. Steel and reinforced concrete are ductile materials that are linear elastic, so the behavior of buildings constructed of these materials can be analyzed by analytic or computational methods. The analysis of buildings made of brittle, unreinforced materials, such as stone, brick, or adobe, can be carried out while the buildings are in the elastic range, before they are damaged. After cracks have formed, analysis becomes extremely difficult, even using modern, advanced-computational capabilities.

A conceptual revolution in seismic design occurred in the 1960s, when engineers developed the concept of *ductile design*. This confers on a structural system the ability to continue to support gravity loads and reversing seismic loads after the building materials have yielded. Prior to this development, the essential approach to seismic design was to provide strength to resist the lateral loads in the structure. Ductile-design approaches have not abandoned strength concepts but have been augmented by reinforcement and connection details, so that elements have the capacity to transmit loads even after they have been damaged. In its simplest form, the term *ductility* has come to mean the ratio of the displacement at which failure occurs (the inability to continue supporting vertical and horizontal loads) to the displacement at which yielding occurs (permanent deformation). Steel and reinforced concrete are characterized as highly ductile materials when the reinforcing materials are used in sufficient quantities and are oriented properly. Brittle materials (e.g., masonry, fired brick, tile, glass, and unreinforced concrete) have high compressive strengths but low ductility, unless reinforced. Unreinforced adobe has low material ductility coupled with low compressive strength; this is generally given as the reason for its poor seismic performance.

The two standard criteria for typical seismic design are (a) to design the structure to remain elastic during moderate to major seismic events; and (b) to design the individual elements and connections of the structure to perform in a ductile manner and retain their strength during major seismic events. The design of the structure in the postelastic phase is not explicitly analyzed. Criteria for the design of concrete and steel construction are based on a combination of field experience and laboratory experimentation.

The Unique Character of Adobe Buildings

The fundamentals of adobe's postelastic behavior are entirely different from those of ductile building materials because adobe is a brittle material. Once a typical unreinforced adobe wall has cracked, the tensile strength of the wall is completely lost, but the wall can still remain standing and can carry vertical loads as long as it remains upright and stable. Cracks in adobe walls may result from seismic forces, from settlement of the foundation, or from internal loads, such as roof beams. Typically, historic adobe walls are quite thick and therefore difficult to destabilize even when they are severely cracked. Support provided at the tops of the walls by a roof system may add additional stability to the walls, especially when the roof system is anchored to the walls. In many adobe buildings, the wall slenderness (height-to-thickness) ratio may be less than 5 and the walls can be 1.2 to 1.5 meters (4 to 5 feet) thick, both of which make wall overturning unlikely. Retrofitting techniques can be used to improve the structural stability of walls and to reduce the differential displacements of the cracked sections of the structure.

Many seismic retrofits of adobe buildings attempt to strengthen adobe walls by addition of ductile, reinforcing elements that allow the wall elements to maintain strength during severe seismic activity. One example is the replacement of the center of an adobe wall with reinforced concrete at the Sonoma Barracks, Sonoma State Historic Park. Such a design is based on the requirement that the wall elements retain strength and ductility and primarily uses elastic design criteria. Reinforced concrete cores have also been placed in the center sections of adobe walls at Petaluma Adobe State Historic Park, in Sonoma County. Cages of concrete beams, grade beams, and reinforced concrete columns have been used at the Plaza Hotel, San Juan Bautista; the Cooper-Molera Adobe, Monterey; and Mission La Purisima Concepcion, Lompoc. However, these types of seismic retrofits are expensive and more intrusive than permitted by conservation standards. In addition, the combination of concrete and brittle adobe may result in problems of material compatibility that will only be realized many years after the original retrofit. In some respects, the building can be considered to be a concrete column and beam structure with adobe brick infill, which represents a significant loss of authenticity.

Reinforced concrete bond beams, placed at the tops of walls, below the roof, are often recommended for the upgrading of existing adobe buildings (California State Historical Building Code; see appendix C). The function of bond beams is to provide lateral support and continuity at the tops of the walls. But the installation of such beams usually requires removal of the roof system, an invasive and destructive procedure. Furthermore, the design of bond beams is often based on elastic design criteria, which usually results in a very stiff beam. After cracks in the adobe walls develop during an earthquake, the stiffness of the bond beam may exceed the stiffness of the walls by two or three orders of magnitude. Adobe walls have pulled out from underneath bond beams during earthquakes due to the difference in stiffness between the bond beam and the cracked wall sections and the lack of a positive connection between the bond beam and the adobe walls. Nonetheless, if the roof system is already slated for removal or for replacement, installation of a *properly anchored* bond beam may well be an appropriate retrofit option.

In the last two decades, many engineering design solutions have been directed toward methods that are less invasive yet structurally effective. Some types of retrofits include steel straps, steel angle bond beams, steel rod crossties at bond beam level, independent steel frame, earth anchors, and plywood diaphragms. A detailed discussion of these approaches in the context of recent seismic retrofit solutions for specific historic adobes is given in Thiel et al. 1991.

Seismic upgrading of existing hazardous buildings is focused on providing maximum life safety to occupants, not on limitation of damage to the buildings. To date, the development of seismic upgrading practices has been directed toward the stabilization of multistory, unreinforced brick masonry (URM) buildings, a ubiquitous building type that is generally regarded as posing the greatest life-safety hazard of all widely used modern building types in the United States. URM structures, on first examination, might be considered to be very similar to adobe buildings, in that walls are built up by stacking bricks and mortar. Yet adobe bricks and mud mortar are much weaker materials than fired brick and cement mortar; therefore, crack damage occurs at much lower levels of earthquake ground motion. More important, however, is the fact that the walls of adobe buildings typically have a much smaller height-to-thickness ratio than the walls of brick buildings. These factors combine to result in a significant difference in the stability problems between adobe and relatively thin-walled brick buildings. Such differences should be recognized and taken into consideration when designing seismic retrofit approaches for the two types of URM buildings.

Structural stability is a fundamental requirement for the adequate performance of adobe buildings during major earthquakes and an important factor when designing appropriate retrofit measures. The massive walls of adobe buildings will crack during moderate to major earthquakes because adobe walls are brittle and adobe is a low-strength material. Seismic ground accelerations act on the massive walls to create large inertial forces that the low-strength adobe material is unable to resist. After cracks have developed, it is essential for the stability of the structure as a whole that the cracked blocks remain in place and able to carry the vertical loads.

A stability-based approach to seismic retrofitting is one that attempts to capitalize on adobe's very favorable postcracking energy-dissipation characteristics and minimize severe structural damage by limiting relative displacements between adjacent cracked blocks. The results of the investigations carried out during the Getty Seismic Adobe Project (GSAP) have shown that a stability-based approach to retrofitting historic adobe buildings can be a most effective method of providing for life safety and of limiting the amount of damage during moderate to severe earthquakes. The purpose of such an approach is to prevent severe structural damage that results in wall collapse. Properly applied, it recognizes the limitations of adobe while taking advantage of the beneficial, inherent structural characteristics of adobe buildings. Thick adobe walls are inherently stable and have great potential for absorbing energy. These characteristics can be greatly enhanced by the application of a number of relatively simple seismic stabilization techniques.

Stability versus Strength

Two fundamental design approaches can be taken to improve the earthquake performance of adobe buildings. *Strength-based design* relies on improving the strength of the adobe material and wall connections and changing the overall structural configuration. This could consist of the addition of shear walls or diaphragms. It assumes the elastic behavior of the building and focuses on traditional means for delaying cracking. *Stability-based design* is concerned with the overall performance of the building and with assuring structural stability during the postelastic, postyielding phase. Stability-based design features can reduce the potential for severe structural damage and collapse after yielding has occurred.

The conventional engineering approach to seismic retrofitting is strength based; that is, structural elements are provided that have sufficient strength to resist the forces generated by the elastic response of the building during a design-level earthquake (the maximum earthquake level the building can endure and still exhibit an elastic, or reversible, response). It is understood that the forces generated during major seismic events can exceed those generated during the design-level event. However, it is also assumed that the nonlinear deformations of the material and connections have sufficient ductility to dissipate the additional energy during a major earthquake.

The second approach—a stability-based design—specifically addresses the postelastic (postdamage) response of the building. This approach requires an understanding of the dynamic characteristics of a damaged adobe structure and the application of techniques that prevent severe damage or collapse. This approach considers severe building damage and the stability, with regard to collapse, of cracked building walls.

These two design strategies are not mutually exclusive: the strength-based approach addresses the elastic behavior of the structure, while the stability-based approach addresses the postelastic performance. In fact, the two approaches can be complementary.

The application of a strength-based analysis alone is not sufficient for determining the performance of thick-walled adobe buildings.
The sole use of an elastic approach can be justified only when there is a
known relationship between the level at which yielding first occurs and
the level at which the structure collapses. In the case of thick-walled
adobe construction, there is no clear relationship between these two
events. Some measures that are designed to improve the elastic behavior
of a building may have little or no effect on structural stability during
major seismic events. Yet stability-based retrofitting measures, which
may have little effect on the initiation or prevention of minor cracks,
may have a significant impact on the development of severe damage and
on preventing collapse.

Figure 4.1 shows a generalized graphic representation of the
differences in the seismic damage behavior of a building retrofitted using
strength-based and stability-based approaches. The horizontal axis is an
increasing function of earthquake intensity and the vertical axis is the
damage index, a qualitative measure of structural damage. Line *ABC*
represents the performance of an unretrofitted building in its original
condition; line *DEF* represents the damageability of the structure with a
strength-based retrofit; and line *GHI* represents the damageability of the
same structure with a stability-based retrofit. An adobe structure is not
damaged until a threshold earthquake intensity is exceeded—point *A*.
Damage then progresses to point *B* with increasing intensity, and then
rapidly to collapse (point *C*) with relatively small increments in intensity.
This line is typical of the behavior of brittle materials. However, many
adobes have structural characteristics (e.g., thick walls) that cause them
to behave less catastrophically.

A classic strength-based retrofit tends to displace the point of
initial damage, *D*, to a higher seismic intensity, after which damage progresses until a critical point, *E*, is reached, above which damage is no
longer repairable. Once the strengthened additions to individual structural elements and connections fail, they have little beneficial effect on
the overall performance of the structure. Beyond *E*, the behavior of the
structure as a whole becomes dominant, and collapse occurs at *F* for
small increments in intensity. For a thin-walled structure, large blocks of
cracked adobe are free to move and are not constrained by other elements of the structure.

Damage to the stability-based retrofitted structure is initiated
at a point close to that of the unretrofitted structure, point *G*, since no
attempt has been made to prevent initial cracking. This strategy uses
nonlinear behavior to advantage and focuses on displacement constraints
rather than strength improvements. The yielding of material and connections progresses to point *H*, where overall behavior begins to dominate
performance. Here the stabilization retrofit elements are engaged, and
the structure exhibits a modestly increasing rate of repairable damage
and resists collapse at relatively high earthquake intensities, *I*.

While strengthening yields distinctly better damage control at
lower intensities, stability-based retrofitting may be the only practical
way to achieve the life-safety objective of preventing collapse. Of course,
implementing a combination of these approaches would be ideal, with

Figure 4.1
Plot of damage-progression index versus earthquake severity for unretrofitted structures (*ABC*) and for stability-based (*GHI*) and strength-based (*DEF*) retrofitted structures.

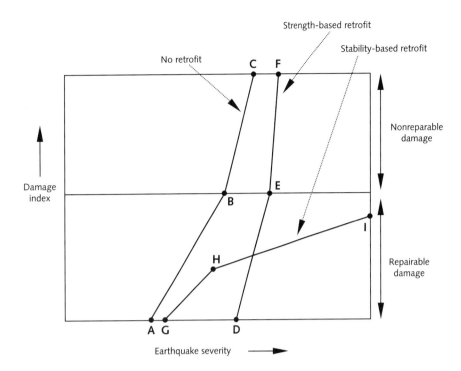

strengthening approaches delaying the initiation of damage, and stability-based retrofitting limiting extensive damage and preventing collapse.

Strength-based design

A strength-based analysis or design procedure uses analytical techniques in which calculations of the resistance of the structure are based on the elastic properties of the material. The dynamic character of earthquake ground motions is most often replaced by an equivalent static force. With a strength-based approach, the assumed design forces are always substantially less (by a factor of 5–10) than the forces that may be expected in major seismic events at a specific site. It is assumed that the ductility of the materials and the connections are sufficient to withstand the stresses produced by these larger seismic events. Conventional strength-based design usually addresses only the consequences of extreme deformations by assessing the elastic deformations under larger-than-design loads.

The conclusions of this type of analysis usually indicate that adobe buildings would not perform well during even moderate seismic ground motions, during which the adobe material will fail, and therefore the building itself will fail. Because historic adobe buildings have massive walls and the adobe itself is a low-strength material, the dynamic or equivalent static forces are large and the tensile properties of the material are easily exceeded. While a strength-based analysis can accurately predict when cracks will occur, it cannot provide insight into the postelastic performance of adobe buildings.

Thin-walled masonry structures can fail catastrophically simply due to gravitational effects shortly after cracks have formed in wall sections. Thick-walled adobe structures, however, are capable of sustaining deflections well beyond the elastic limit of the material. The stability of such walls may not be threatened even when wall deflections are more than one hundred times the deflection at the elastic limit of the adobe material.

The structural ductility (not material ductility) of a building system is a critically important characteristic of the seismic design of a building. *Structural ductility* is defined as the capacity of a building to maintain its load-carrying capability and deform safely after the elastic limit of the building material has been exceeded. Thick-walled adobe buildings can exhibit substantial structural ductility even though the building's construction material itself is brittle.

Stability-based design

For adobe buildings, a stability-based design analysis can take advantage of the unique characteristics of the postelastic performance of adobe and the effects of a proposed retrofit system. The walls must be relatively thick, as they are in the vast majority of existing historic adobe structures, for the building to exhibit the ductile performance characteristics required to resist the destructive forces of major earthquakes.

It is often assumed that an unreinforced masonry structure (such as adobe or brick) is safe only while it is largely undamaged, that is, if it has not sustained substantial cracking. The usual analysis assumes that once cracks have developed the materials have lost strength and continuity—and therefore the building is unsafe. However, a thick-walled adobe building is *not* unstable after cracks have fully developed, and the building still retains considerable stability characteristics even in that state. Retrofit measures can greatly enhance overall stability and act to limit the extent of damage in the form of large permanent offsets.

The extent of retrofit intervention required to stabilize an adobe wall is often relatively small and relies on many of the inherent properties of historic adobe construction. The following are some of the important attributes of a stability-based retrofit design:

- **Allows out-of-plane rocking.** Out-of-plane stability of thick adobe walls is not as serious a consideration as is generally assumed by conventional, strength-based methods.
- **Provides out-of-plane restraint at the tops of walls.** Additional restraint at the top of thick adobe walls will greatly increase the out-of-plane stability of cracked blocks.
- **Provides flexible connections between perpendicular walls that tie the walls together.** Perpendicular walls have very different deflection characteristics, so flexible connections are important.
- **Provides ties that resist the relative and permanent displacement of adjacent, cracked blocks.** Very little force is required to greatly reduce both in-plane and out-of-plane block movements during extended seismic excitations.

Performance-Based Design

The current trend in engineering design is to design for multiple, specifically defined levels of performance at different earthquake levels. Building codes historically use a design methodology in which the ultimate failure of a building is an implicit part of the design (Hamburger et al. 1995). Performance-based seismic design of conventional construction uses a variety of modern design methodologies, but for unreinforced adobe construction these techniques are sorely lacking.

The fundamental goal of performance-based design is to predict a building's response accurately during increasing levels of seismic excitation. Since numerical methods for adobe buildings cannot yield accurate results, heavy reliance must be placed on the actual knowledge of the seismic behavior of buildings obtained from either observations in the field or the results of simulations in the laboratory. GSAP research relied on both sources of information for developing and testing the suggested retrofit techniques (see chap. 7). The goal of these guidelines is to extrapolate those field observations and test results for use on other buildings.

Current Building Codes and Design Standards

In California, the prevailing building code for designated historic buildings is the California State Historical Building Code (SHBC; California's State Historical Building Safety Board 1999). The SHBC has specific recommendations for adobe buildings, and the 1998 version of the SHBC allows for the use of the 1994 edition of the Uniform Code for Building Conservation (UCBC) with unreinforced masonry buildings.

The essence of the SHBC and UCBC design methods is simply to specify allowed strength values for adobe buildings and suggested design levels. The SHBC allows 4 pounds per square inch (psi) and the UCBC allows 3 psi. The design loads are as prescribed by the 1994 Uniform Building Code (UBC) for the SHBC. The UCBC design forces are either 10% or 13% of gravity in the most active seismic zone (Zone 4) depending on the occupancy levels. The SHBC limits the wall height-to-thickness ratio to 5 for the first floor and 6 for the second floor or for single-story buildings. The UCBC allows for a height-to-thickness ratio of 8. This acceptable height-to-thickness ratio is based on the results of the GSAP research program. Walls that meet these maximum height-to-thickness ratios do not require additional strengthening measures. The SHBC suggests the use of concrete bond beams or an equivalent design using other materials at the floor and roof levels.

Anchorage forces are not explicitly addressed in the 1998 SHBC, but the UCBC defers to the values assigned by the UBC. No other references are supplied for anchorage of adobe walls. The values derived from UBC calculations result in rather close spacing of anchor bolts. In the moderate- and thick-walled building models tested as part of GSAP, the anchorages were placed at intervals greater than those suggested by the UBC and did not fail (Tolles et al. 2000). However, GSAP research did not explicitly address issues of adobe anchorage and spacing.

These details for the design of adobe buildings under the UCBC or SHBC are values to be used in simple, strength-based design procedures. Conformance with these standards may be misleading (a building may not be safe during major seismic events) or may simply be a distraction from the real issues involved in executing a successful design.

As an example, assume we are designing a simple rectangular adobe building that is 6.1 meters (20 feet) wide and 24.4 meters (80 feet) long, as shown in figure 4.2. The walls have a height-to-thickness ratio of 4. They are very stout, and overturning is unlikely if the condition at the base is good.

Based on a stability analysis, the walls should have nominal anchorage at the tops (approximately 4 feet on center) and the system should be tied together through the flexible, low-strength roof diaphragm or cabling. Vertical center-core rods could be added at regular or selected locations to limit damage during major events. But, with or without the center-core rods, it would be extremely difficult for such a structure to collapse.

Based on a strength-based analysis, the top-of-wall anchors should be closely spaced—approximately 30 to 46 cm on center (12 to 18 inches)—the roof needs strengthening, and the actual shear stresses in the walls would be significantly higher than with a more flexible roof system. The principal concern is to distribute the out-of-plane forces into the in-plane walls, even though the true seismic behavior of this building does not warrant this type of intervention.

The most serious problem for this building may be the southwest corner. The window and door are located very close to this corner, and collapse of the entire corner might occur even with the presence of top-of-wall anchors (fig. 4.3). Simple adherence to the strength-based procedures could lead to a false sense of security.

A proper stability-based analysis would recognize this area as a problem. A map of the predicted and possible crack patterns (see chap. 7) would allow the potential areas of instability to be identified and allow for additional retrofit measures (straps, cables, or center-core elements) to prevent this type of local instability.

The design of historic adobe buildings should use strength-based procedures only as a very general guideline. Of greater importance is the identification of the areas of a building that may be fragile and susceptible to serious damage or collapse.

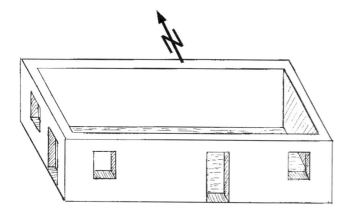

Figure 4.2
Undamaged small adobe building.

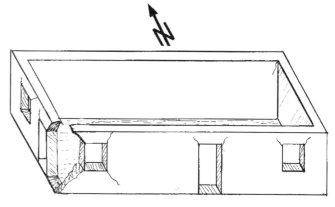

Figure 4.3
Damaged small adobe building.

Chapter 5

Characterization of Earthquake Damage in Historic Adobe Buildings

Documentation of the actual damage resulting from strong earthquakes is essential to understanding how historic adobe buildings do, in fact, behave in earthquakes. While it is true that portions of, or entire, adobe buildings may collapse during strong earthquakes, it is *not* true that adobe buildings are unstable simply because the walls have cracked. This chapter describes the types of damage that can occur and how much damage can be expected in historic adobe buildings during strong ground motions.

Damage Levels in Aseismic Design

The behavior of adobe buildings undergoes significant changes as seismic ground motions increase in magnitude. As long as the building is undamaged, it will respond elastically for a short time. The use of known analytical techniques can approximate this dynamic behavior. If the building walls are cracked, and independent adobe blocks have already formed, then the applicability of standard elastic analysis is questionable, since the adobe material is no longer continuous. After cracks have developed, the stability of the walls depends on gravity, and the Coulomb friction along cracks between block elements becomes important.

Cracking is almost certain to occur during major seismic ground motions as the stresses in the walls exceed the tensile capacity of the adobe material. As cracks develop, the dynamic response characteristics of the structure undergo drastic changes: the fundamental vibration frequency decreases dramatically, and the magnitude of the wall or block displacements can increase by two to three orders of magnitude. Motion along cracks becomes substantial as cracks intersect and independent adobe blocks are formed. Thus the dynamic behavior of such a damaged adobe structure cannot be predicted using available analytical techniques, which are applicable only to modeling the elastic response of an undamaged adobe building.

Elastic behavior

The elastic behavior of most adobe structures is characterized by a relatively high-frequency response and small structural-distortion displacements. Even though the adobe material has a low modulus of elasticity—typically less than 690 MPa (100,000 psi)—compared to that of other building materials, the walls are usually quite thick, have fewer openings, and are therefore relatively stiff. The frequency of the principal mode of vibration of a typical single-story residential building is in the 5–10 hertz range. Larger adobe structures, such as mission church buildings, will have lower fundamental frequencies in the undamaged state, but these frequencies will still be relatively high compared to the vibration frequency of those in a cracked condition.

Initial cracking

Substantial cracks nearly always exist in historic adobe buildings as a result of past earthquake activity, wall slumping, or foundation settlement. Cracked walls are a typical feature of these buildings, and cracks usually develop in areas of high stress concentrations, such as the corners of openings (doors and windows), at the intersections of perpendicular walls, and at the base of walls. Cracks at doors and windows can develop from either out-of-plane (flexure) or in-plane (shear) forces in the walls. Vertical or diagonal cracks at wall intersections occur as a result of a combination of flexural and tensile stresses. The out-of-plane motion of long walls often results in horizontal cracking near the base of the wall. Gravity loads induced by the weight of the wall and the tributary loads largely influence the vertical location of these horizontal cracks. In the massive adobe walls of mission churches, for example, horizontal cracking can occur from 1.5 to 3 m (5–10 ft.) up from the base of the wall.

The performance of an adobe building is substantially affected by the thickness of its walls. Moderate and thick walls are defined here in terms of the slenderness, or wall height-to-thickness, ratio (S_L):

- thick: $S_L < 6$
- moderate: $S_L = 6$–8
- thin: $S_L > 8$

Thin adobe walls may become unstable soon after the initiation of cracks through the wall. However, a thick-walled adobe building is still a long way from losing its stability after the first cracks develop. An adobe building must undergo many changes in dynamic characteristics and sustain much larger displacements than those required for initial cracking before a thick wall approaches instability.

Changes in dynamic behavior

The dynamic characteristics of an adobe building change dramatically as cracks develop. For long, out-of-plane walls, the effective frequency of motion decreases and the dynamic displacements increase. The term *effective frequency* is used to represent the apparent frequency of motion for nonlinear materials. This change in behavior can be demonstrated by an example taken from GSAP experimental data on a shaking-table test of a model adobe building, given in figure 5.1. The first plot (fig. 5.1a) shows out-of-plane wall accelerations; the second plot (fig. 5.1b) shows displacements of a building during ground motion, as a function of time, while damage is developing. About midway through the time history, the

Figure 5.1

Plots of out-of-plane acceleration of the tops of walls, showing decrease as damage develops. The first test shows (a) decrease in wall accelerations with time during shaking and (b) increasing displacements with time. During the next test (c) accelerations are much lower, even though the input shaking-table displacement was about 30% higher.

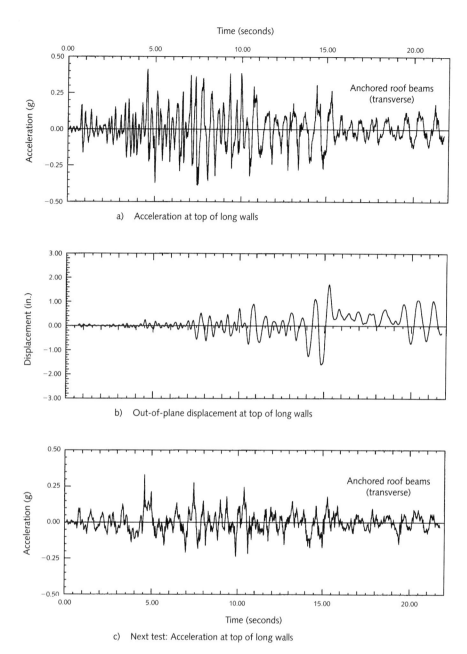

a)   Acceleration at top of long walls

b)   Out-of-plane displacement at top of long walls

c)   Next test: Acceleration at top of long walls

wall accelerations begin to decrease, the fundamental vibration frequency decreases, and the displacements began to grow dramatically. The third plot (fig. 5.1c) shows the wall acceleration in the subsequent test. Even though the input motion is approximately 30% higher than that of the previous test, both the peak wall accelerations and the effective frequency have decreased.

The difference in out-of-plane wall displacements between damaged and undamaged buildings is best shown by a direct comparison of the out-of-plane displacement of a damaged and an undamaged model building during the same test. In the results shown in figure 5.2, the displacement of the damaged building is nearly ten times larger than that of the undamaged structure. Even with the large increase in displacement, the value shown for the damaged wall (fig. 5.2b) is still only one-tenth of that required for overturning.

Figure 5.2

Comparison of out-of-plane displacements during shaking-table test: (a) undamaged wall, and (b) wall after substantial cracking had occurred.

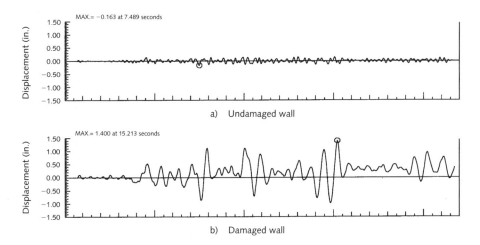

In-plane wall displacements and accelerations undergo less dramatic changes as cracks develop, and they are not usually a threat to the stability of an adobe structure. As cracks occur and blocks slide, friction along adjacent cracked blocks limits the in-plane wall displacements much more than it limits displacements in the out-of-plane direction. However, when diagonal cracks develop at a building corner, progressive in-plane or out-of-plane failure may result.

Moderate-to-heavy damage and collapse potential

As damage to an adobe building progresses, crack sizes increase during reversing cycles of ground motion, and the building's effective frequency continues to decrease. When the crack pattern is fully developed, each wall becomes an assemblage of irregularly shaped blocks of wall segments. These blocks, referred to as a *cracked wall section*, may be restricted to a portion of the wall height or may extend from one floor line to the next. A cracked wall section may also extend from one interior wall to the next or from opening to opening. Even though the building may have fully developed cracks and independent blocks of adobe may have formed from intersection of these cracks, the structure is still likely to be able to endure considerable ground motion without becoming unstable.

Out-of-plane wall displacements pose the greatest collapse threat to an adobe building. The three primary factors that affect the out-of-plane stability (overturning) of badly cracked walls are (1) the absolute thickness and slenderness (height-to-thickness) ratio of the wall; (2) restraints that may limit the deflection at the top (connection at the floor or roof line) or the sides (perpendicular walls), or between blocks formed in the wall; and (3) added gravity loads from roof or floor framing. Vertical cracks may develop such that perpendicular walls provide little or no stabilizing effects. It should be noted that very little restraint (force) is required at the top of a block to reduce substantially the overturning potential of a cracked wall section.

Non-load-bearing walls, even when shorter than the bearing walls, are usually the first to collapse. This occurs because in most historic adobe buildings often little or no restraint is provided by roof or floor connections and there are no additional tributary loads. The hazard is particularly great for gable-end walls due to their larger slenderness (height-to-thickness) ratios and minimal connections to the roof and floor systems.

Load-bearing walls may also collapse, and this is likely to be catastrophic from both conservation and life-safety perspectives. For thick adobe walls with fully developed cracks, the length of the wall will have little effect on the overturning potential of that wall. A wall 30 m (98 ft.) long with developed cracks at the base and sides may present no greater hazard of collapse than a wall 4 m (13 ft.) long with similarly developed cracks. The primary factors affecting collapse of a bearing wall are its absolute thickness, its slenderness ratio, and the degree of restraint at the top. The longitudinal dimension of a wall, or an independent cracked block, may have little effect on the wall's potential for overturning and collapse, unless the top of the wall is anchored to cross walls at the floor or roof level. Adequate connection between the walls and either the roof or flooring system is essential to prevent overturning. Inadequate bearing of roof or floor beams and the lack of a positive connection can allow a load-bearing wall to move progressively out from under the beams.

In-plane shear damage to walls will become more severe during strong seismic events, and diagonal cracks may develop in sections of walls that have no openings. The movement of wall blocks may be increased as the shaking continues, and cracked wall sections near the ends of walls will be susceptible to permanent offsets along diagonal cracks.

Evaluating the Severity of Earthquake Damage

This section defines, in general terms, the relative level and location of damage that can occur in adobe buildings following an earthquake and discusses the correlation between peak ground acceleration and damage states.

Damage states
For describing and comparing the relative damage levels sustained by buildings following an earthquake, it is useful to have a standard guide that describes the escalation of the severity of damage. Such a set of standardized damage states has been developed by the Earthquake Engineering Research Institute (EERI 1994). Table 5.1 contains a description of each damage state and a corresponding description of damage in historic adobe buildings.

Damage typologies
The following subsections include descriptions, figures, and photographs of the damage types observed in historic adobe buildings. The typical damage types are illustrated in figure 5.3, and a more complete listing is presented in table 5.2.

It is important to understand the relative severities of the various types of damage as they relate to life safety and the protection of

Table 5.1

Standardized damage states

Damage state	EERI description	Comments on damage to historic adobe buildings
A None	No damage, but contents could be shifted. Only incidental hazard.	No damage or evidence of new cracking.
B Slight	Minor damage to nonstructural elements. Building may be temporarily closed but probably could be reopened after cleanup in less than one week.[a] Only incidental hazard.	Preexisting cracks have opened slightly. New hairline cracking may have begun to develop at the corners of doors and windows or the intersection of perpendicular walls.
C Moderate	Primarily nonstructural damage but there could be minor, nonthreatening structural damage. Building probably closed for 2–12 weeks.[a]	Cracking damage throughout the building. Cracks at the expected locations (openings, wall intersections, slippage between framing and walls). Offsets at cracks are small. None of the wall sections are unstable.
D Extensive	Extensive structural and nonstructural damage. Long-term closure should be expected, due either to amount of repair work or uncertainty of economic feasibility of repair. Localized, life-threatening situations would be common.	Extensive crack damage throughout the building. Crack offsets are large in many areas. Cracked wall sections are unstable. Vertical support for the floor and roof framing is hazardous.
E Complete	Complete collapse or damage that is not economically repairable. Life-threatening situations in every building in this category.	Very extensive damage. Collapse or partial collapse of much of the structure. Due to extensive wall collapse, repair of the building requires reconstruction of many walls.

[a] Length of time is difficult to assign because it is largely dependent on the size of the building and the process used for repairs. Repairs to historic buildings should be undertaken in a much more deliberate manner than is typical in the repair of a more modern building.

historic building fabric. By doing so, priorities for stabilization, repairs, and/or seismic retrofits can be established for each type of damage. If a particular damaged area or component of a building is likely to degrade rapidly if not repaired, then that damaged element assumes a higher priority than others that are not likely to deteriorate. If damage to a major structural element, such as a roof or an entire wall, increases the susceptibility to collapse, then a high priority is assigned because of the threat to life safety. If damage that could result in the loss of a major feature, such as a wall, compromises the historic integrity of the entire structure, then it is more critical than damage that would result in partial failure, but no loss.

Table 5.2 provides details of the life-safety and historic-fabric concerns for each of the damage types. As noted in the table, some damage types are usually not serious, but they may become serious if the structure is subjected to greater loads, loads of longer duration, or repeated earthquakes—particularly when no remedial repairs are carried out.

In most situations, different types of damage do not act independently but rather in combination. In fact, several of the damage types are actually caused by other types. In some cases, the specific relation-

Figure 5.3
Typical damage modes observed in historic adobe buildings after the 1994 Northridge earthquake.

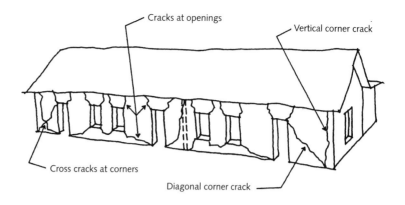

Cracks at openings

Vertical corner crack

Cross cracks at corners

Diagonal corner crack

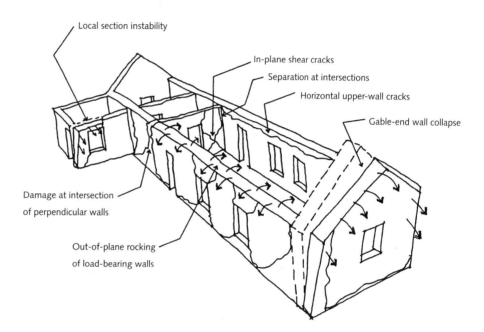

Local section instability

In-plane shear cracks

Separation at intersections

Horizontal upper-wall cracks

Gable-end wall collapse

Damage at intersection
of perpendicular walls

Out-of-plane rocking
of load-bearing walls

ships among different damage types are simple, while in others they may be extremely complex.

Out-of-plane wall damage

Adobe walls are very susceptible to cracking from flexural stresses caused by out-of-plane ground motions. The cracks caused by out-of-plane flexure usually occur in a wall between two transverse walls. The cracks often start at each intersection, extend downward vertically or diagonally to the base of the wall, and then extend horizontally along its length. The wall rocks back and forth out of plane, rotating about the horizontal crack at the base. Cracks due to out-of-plane motions are typically the first type of damage to develop in adobe buildings. Out-of-plane cracks develop in an undamaged adobe wall when peak ground accelerations reach approximately 0.2g.

Although wall cracks that result from out-of-plane forces occur readily, the extent of damage is often not particularly severe, as long as the wall is prevented from overturning. The principal factors that affect the out-of-plane stability of adobe walls are as follows:

Table 5.2

Historic adobe earthquake damage typologies and their effect on life safety and historic fabric

Type	Description	Life safety and historic fabric concerns
Out-of-plane damage Gable-end wall failure	Gable-end walls suffer severe cracking that often leads to instability. They are tall, poorly attached to the building, have large slenderness (height-to thickness) ratios, and carry no vertical loads. These walls are highly susceptible to collapse.	Collapse of gable-end walls is a serious life-safety threat and causes extensive loss of historic fabric.
Out-of-plane damage Flexural cracks and collapse	Flexural cracks begin as vertical cracks at transverse walls, extend downward vertically or diagonally to the base of the wall, and extend horizontally to the next perpendicular wall. *The existence of cracks does not necessarily mean that a wall is unstable.* Walls can rock without becoming unstable. After cracks have developed, the out-of-plane stability of a wall is dependent on the slenderness ratio, connection to the structure, vertical loads, and the condition of the wall at its base.	When walls only develop cracks and are stabilized at the top to prevent over-turning, this damage type is not severe. Many load-bearing walls in extensively damaged adobe buildings were stable throughout the Northridge earthquake. In the case of overturning, the life-safety danger is serious because not only do the walls collapse but the roof or ceiling structure may also collapse.
Out-of-plane damage Mid-height cracks	Long, tall, and slender single-wythe walls, or long, tall double-wythe walls with no header courses interconnecting the wythes are susceptible to mid-height horizontal cracking from out-of-plane ground motion.	Damage represented by mid-height horizontal cracking is not serious in and of itself. However, the potential for much greater damage is significant. During further ground shaking, out-of-plane movement of the wall could cause the upper or lower sections of the wall to become unstable and collapse, thus creating a life-safety threat.
In-plane damage	Classic X-shaped or simple diagonal cracks are caused by in-plane shear forces.	In-plane shear cracks generally do not constitute a life-safety hazard. Nevertheless, this type of damage can cause extensive damage to the walls and the attached plaster, which may be historic. When large horizontal and vertical offsets occur at these cracks, repair costs may be significant and a loss of historical integrity can result.
Corner damage Vertical	Vertical cracks can develop at corners in one or both planes of intersecting walls.	Life-safety hazard is minimal. The collapse of an entire corner can occur when vertical cracks occur in both planes of a corner, resulting in loss of historic fabric and a costly repair.
Corner damage Diagonal	Diagonal cracks that extend diagonally from the bottom to the top of a wall at a corner may be caused by in-plane shear forces or out-of-plane flexural forces.	Life-safety hazard is minimal. Slippage can occur along diagonal cracks that slant downward toward a corner. If much vertical slippage occurs, the wall may be very difficult to repair, compromising historical integrity.
Corner damage Cross	A diagonal crack extending from the bottom corner can combine with a diagonal crack from the top corner forming a wedge-shaped section.	Life-safety hazard is minimal. A complex pattern of cracks can lead to significant offsets of sections of the walls. Damage may be difficult to repair if these offsets occur, compromising historical integrity.
Cracks at openings	Cracks often begin at the tops of doors and openings and propagate upward vertically or at a diagonal. Cracks can also develop at the lower corners of windows. These cracks may be caused by in-plane or out-of-plane motion.	Life-safety hazard is minimal. The cracks that occur at the tops and bottoms of openings are typically not severe except as they affect the plaster over and around the cracks, which may be historic.

Table 5.2
continued

Type	Description	Life safety and historic fabric concerns
Damage at intersection of perpendicular walls	Perpendicular walls can separate from each other and become damaged by pounding against each other.	Life-safety hazard is minimal, unless other problems occur as a result of this damage. Damage to historic fabric is minimal, unless historic renderings spall.
Slippage between walls and wood framing	Roof, ceiling, and floor framing often slips at the interface with the adobe walls. Wood framing is often not attached or is inadequately attached to the adobe walls in historic adobe buildings.	If the slippage between the walls and wood framing is not large, then the life-safety hazard is minimal, but it may still be costly to repair. If the slippage is large, it may indicate the walls were approaching instability, which presents a very hazardous life-safety condition. Normally, historic fabric is only slightly compromised.
Damage at wall or tie-rod anchorage	Crack damage often propagates from structural anchorage and crossties. It is difficult to avoid stress concentrations at these locations, and this generally leads to cracks and other damage such as crushing of material.	Life-safety hazard is minimal, unless the local damage leads to other more significant problems. Damage to historic fabric is localized.
Local section instability	Local wall sections can become unstable as the result of cracks that develop at corners of buildings and/or window and door openings.	In the immediate area, life-safety hazards and loss of historic fabric may be significant.
Horizontal upper-wall cracks	Horizontal cracks may develop near the tops of walls when there is a bond beam or the roof is anchored to it. These cracks are caused by the combination of horizontal forces and the small vertical compressive stresses near the top of the wall.	Life-safety hazard is minimal. These cracks occur when there are bond beams or if the roof is anchored to the walls. If the bond beams are not anchored to the walls, they may slip. Otherwise, there is usually only a horizontal crack at the interface, which is not particularly significant.
Moisture-damage contributions to instability	Moisture damage at the base of a wall can result in wall instability. In some cases, the wall may collapse out of plane because one side of the wall has been weakened or eroded. In other cases, saturation or repeated wet-dry cycles can weaken the lower adobe walls, causing weakened slip-planes at the base, along which the wall can slip and collapse.	Significant potential life-safety hazard. Moisture damage at the base of a wall can lead to instability and collapse of a wall that would have otherwise been stable. Restraint at the top of the wall will have little effect on stabilization.

- wall thickness and the slenderness ratio (S_L);
- the connection between the walls and the roof and/or floor system;
- whether the wall is load-bearing or non-load-bearing;
- the distance between intersecting walls; and
- the condition of the base of the wall.

The slenderness ratio of a wall is a fundamental indication of its stability (resistance to overturning). It is virtually impossible to overturn a thick wall ($S_L < 6$); it will slip horizontally at the base before it will overturn. On the other hand, slender walls ($S_L > 8$) are very susceptible to overturning or possibly buckling at mid-height.

The presence of connections between the walls and the roof and/or floor systems can greatly improve the out-of-plane stability of a wall. It is not necessary for the floor or roof to constitute a complete diaphragm system for these connections to improve the out-of-plane stability significantly. A wood bond beam or partial plywood diaphragm may be sufficient to stabilize the out-of-plane motions of walls. Even anchoring a wall to a roof or floor system without strengthening the diaphragm can positively affect out-of-plane stability. Vertical loads at the tops of thicker, load-bearing walls also act to stabilize the walls. As the wall rocks out of plane, the load shifts to the edge of the wall that is rocking upward and resists the overturning by bearing down on the raised corner.

The condition of the base of an adobe wall may also affect its out-of-plane stability. The following conditions lead to out-of-plane instability or increase susceptibility to overturning: basal erosion, which reduces the bearing area; excessive moisture content, which reduces the strength; and repeated wet-dry cycles, which may also weaken the adobe.

The collapse of any wall is obviously a serious failure—one that results in a loss of historic fabric and carries a high reconstruction cost and a grave risk of serious injury.

Gable-end wall collapse

Overturning—the principal damage to gable-end walls—is a special case of out-of-plane failure that needs specific discussion because these walls are typically the elements most susceptible to damage in historic adobe buildings. Gable-end walls are tall and thin, non-load-bearing, and usually not well connected to the structure at the floor, attic, or roof level. Their overturning is caused by ground motions that are perpendicular (out of plane) to the walls. Instability problems can also result from in-plane ground motions when sections of the wall slip along diagonal cracks and then become unstable out of plane, especially at corners.

The most severe damage results in the collapse of an entire wall, which destroys extensive amounts of historic fabric and poses a serious threat to life safety (fig. 5.4a). Out-of-plane flexure may or may not be severe, depending on the extent of cracking and permanent displacement. Of particular concern is a section of wall that becomes independent of the rest of the structure by virtue of the crack pattern (fig. 5.4b). The instability of such a section is considered serious because,

Figure 5.4
Gable-end wall collapse: (a) overturning at base of wall, and (b) mid-height collapse.

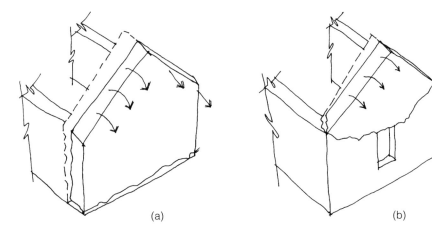

(a)   (b)

without restraint, it may collapse following even moderate additional ground motions.

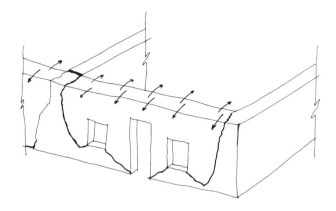

Figure 5.5
Out-of-plane flexure of load-bearing wall.

Figure 5.6
Overturned, unsupported garden wall ($S_L = 5$). Mission San Fernando after 1994 Northridge earthquake.

Out-of-plane flexure cracks and collapse

Out-of-plane flexural cracking is one of the first crack types to appear in an adobe building during a seismic event. This damage type and the associated rocking motion are illustrated in figure 5.5. Freestanding walls, such as garden walls, are most vulnerable to overturning because there is usually no horizontal support along their length, such as that provided by cross walls or roof or floor systems. Such a wall, a garden wall on the north side of the *convento* at the San Fernando Mission, overturned during the Northridge earthquake (fig. 5.6). Although this wall was constructed of stabilized adobe bricks and had a slenderness ratio of 5, it could not withstand the severe out-of-plane rocking to which it was subjected.

Mid-height, out-of-plane flexural damage

For the most part, historic adobe buildings are not susceptible to mid-height, out-of-plane flexural damage because the walls are usually thick and have small slenderness ratios. However, horizontal cracks may develop when load-bearing walls are long and the top of the wall is restrained by a bond beam or a connection to a roof or ceiling system (fig. 5.7). This type of damage and potential failure mechanism is usually observed only in thin-walled ($S_L > 8$) masonry buildings.

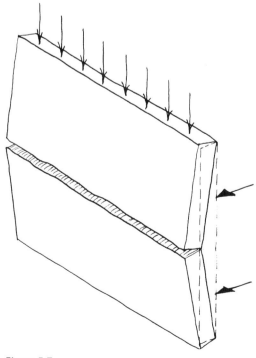

Figure 5.7
Mid-height, out-of-plane flexure damage.

In-plane shear cracks

Diagonal cracks (figs. 5.8a, b) are typical results of in-plane shear forces. The cracks are caused by horizontal forces in the plane of the wall that produce tensile stresses at an angle of approximately 45 degrees to the horizontal. Such X-shaped cracks occur when the sequence of ground motions generates shear forces that act first in one direction and then in the opposite direction (fig. 5.8c). These cracks often occur in walls or piers between window openings.

The severity of in-plane cracks is judged by the extent of the permanent displacement (offset) that occurs between the adjacent wall sections or blocks after ground shaking ends. More severe damage to the structure may occur when an in-plane horizontal offset occurs in combination with a vertical displacement, that is, when the crack pattern follows a more direct diagonal line and does not "stair-step" along mortar joints. Diagonal shear cracks can cause extensive damage during prolonged ground motions because gravity is constantly working in combination with earthquake forces to exacerbate the damage.

In-plane shear cracking, damage at wall and tie-rod anchorages, and horizontal cracks are relatively low-risk damage types.

Figure 5.8

Illustrations show (a) drawing of X-shaped shear cracks in an interior wall; (b) typical X pattern (Leonis Adobe, Calabasas, Calif.); and (c) how X-shaped cracks result from a combination of shear cracks caused by alternate ground motions in opposite directions.

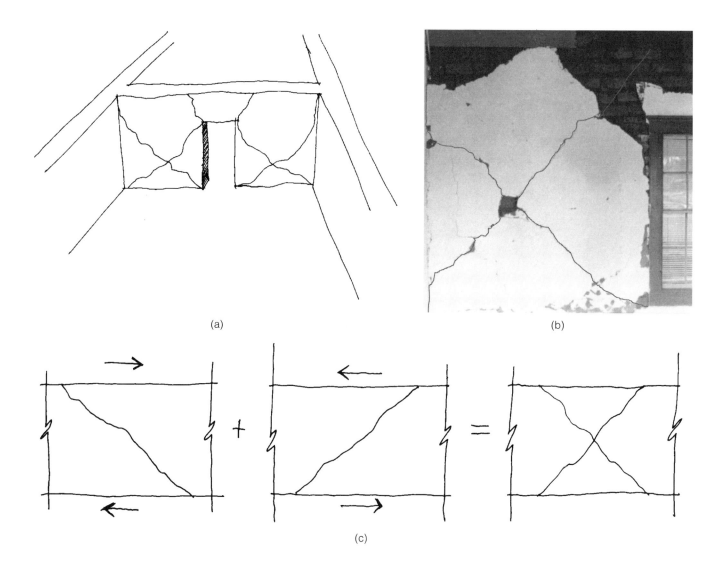

(a)

(b)

(c)

However, while in-plane shear is not considered hazardous from the perspective of life safety, it is often costly in terms of loss to historic fabric. In-plane shear cracks often cause severe damage to plasters and stuccos that may be of historic importance, such as those decorated with murals.

Corner damage

Damage often occurs at the corners of buildings due to the stress concentrations that occur at the intersection of perpendicular walls. Instability of corner sections often occurs because the two walls at the corner are unrestrained and therefore the corner section is free to collapse outward and away from the building.

Vertical cracks at corners

Vertical cracks often develop at corners during the interaction of perpendicular walls and are caused by flexure and tension due to out-of-plane movements. This type of damage can be particularly severe when vertical cracks occur on both faces, allowing collapse of the wall section at the corner (fig. 5.9).

Diagonal cracking at corners

In-plane shear forces cause diagonal cracks that start at the top of a wall and extend downward to the corner. This type of crack results in a wall section that can move laterally and downward during extended ground motions. Damage of this type is difficult to repair and may require reconstruction. Illustrations of this damage type are shown in figure 5.10.

Combinations with other cracks or preexisting damage

A combination of diagonal and vertical cracks can result in an adobe wall that is severely fractured, and several sections of the wall may be susceptible to large offsets or collapse. An example of a wall section that is highly vulnerable to serious damage is illustrated in figure 5.11 (see also fig. 5.9a). The diagonal cracking at that location allows the cracked wall sections freedom to move outward. Corners

Figure 5.9
Illustrations showing (a) how vertical cracks at corner can lead to instability of intersection, and (b) example of corner collapse (Sepulveda Adobe, Calabasas, Calif.).

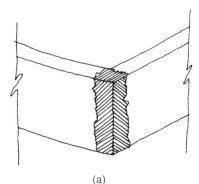

(a)   (b)

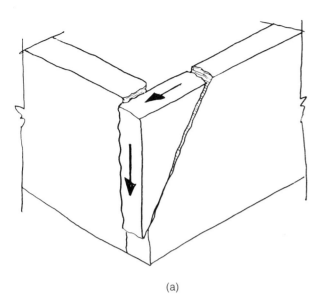

(a)

(b)

Figure 5.10
Corner cracks: (a) illustration of vertical downward and horizontal displacement of a corner wall section, and (b) example of displaced wall section (Leonis Adobe).

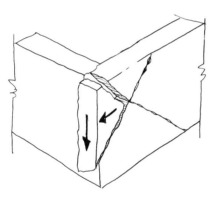

Figure 5.11
Illustration showing how combination of shear and flexural cracks can result in corner displacement or collapse.

may be more susceptible to collapse if vertical cracks develop and the base of the wall has already been weakened by previous moisture damage.

Cracks at openings

Cracks occur at window and door openings more often than at any other location in a building. In addition to earthquakes, foundation settlement and slumping due to moisture intrusion at the base can also cause cracking. Cracks at openings develop because stress concentrations are high at these locations and because of the physical incompatibility of the adobe and the wood lintels. Cracks start at the top or bottom corners of openings and extend diagonally or vertically to the tops of the walls, as illustrated in figure 5.12.

Cracks at openings are not necessarily indicative of severe damage. Wall sections on either side of openings usually prevent these cracks from developing into large offsets. However, in some cases, these cracks result in small cracked wall sections over the openings that can become dislodged and could represent a life-safety hazard.

Intersection of perpendicular walls

Damage often occurs at the intersection of perpendicular walls. One wall can rock out of plane while the perpendicular in-plane wall remains very

Figure 5.12
Illustration of cracks originating at stress concentration locations: (a) cracks appearing first at upper corners of window opening, followed by lower corner cracks; and (b) cracks at upper corners of door opening.

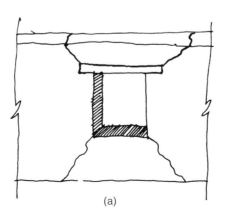

(a)

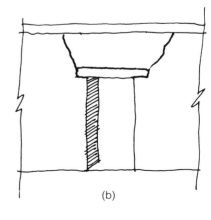

(b)

stiff. Damage at these locations is inevitable during large ground motions and can result in the development of gaps between the in-plane and out-of-plane walls (fig. 5.13a) or in vertical cracks in the out-of-plane wall (fig. 5.13b). Damage may be significant when large cracks form and associated damage occurs to the roof or ceiling framing. Anchorage to the horizontal framing system or other continuity elements can greatly reduce the severity of this type of damage.

Damage at the intersection of perpendicular walls is normally not serious from a life-safety perspective. However, in the same way that corner damage occurs, adjacent walls can become isolated and behave as freestanding walls. When they reach this state, the possibility of collapse or overturning is greatly increased, and a serious life-safety threat can arise. In addition, if significant permanent offsets occur, repair may be difficult and expensive.

Slippage between adobe walls and roof, ceiling, or floor framing

Slippage often occurs between the horizontal framing of second floors, ceilings, or roofs and adobe walls. In typical historic adobe buildings, there is little attachment between the walls and the framing. Second-floor or ceiling joists are usually set into pockets in the tops of walls. Ceiling or roof-framing members are often set directly on the tops of walls, with or without wall plates. As a result, permanent offsets between ceiling framing and the adobe wall (fig. 5.14) are quite commonly observed, and failures of this type can range from cosmetic to severe. The adobe walls may slip out from under the framing, which could lead to collapse of the wall, ceiling, and roof. This condition has also been observed in newer

Figure 5.13
Illustrations showing (a) how separation can occur between in-plane and out-of-plane walls, and (b) how vertical cracks develop in out-of-plane walls at the intersection with perpendicular, in-plane walls.

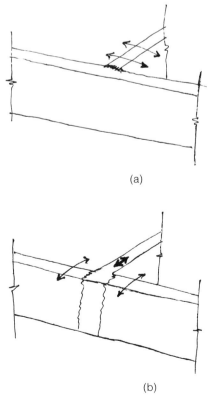

(a)

(b)

Figure 5.14
Offset between *tapanco* floor joists and load-bearing wall (Del Valle Adobe, Rancho Camulos Museum, Piru, Calif.).

adobe buildings constructed using concrete bond beams where, as a result of the lack of mechanical attachment between the adobe wall and the bond beam, the wall pulled out from beneath the stiff bond beam.

Slippage between walls and ceiling or roof framing is normally not serious in terms of impact on historic fabric. In some cases, the bearing area of ceiling joists on the wall is inadequate, and slippage can create a serious life-safety threat. However, unless other parts of the structure fail at the same time, excessive slippage is not likely to cause the roof to collapse. Of course, if both walls that support the roof have moved outward, then the situation is extremely critical, and catastrophic collapse of the entire roof can occur.

Figure 5.15
Anchorage failure of steel tie-rod; anchor plate has pulled into the wall.

Damage at wall anchorage

Wall anchors (tie-rods) are intended to hold a wall snugly against a perpendicular wall or diaphragm. It is common for wall anchors to have been installed following an earthquake or settlement damage. Subsequent damage to walls can then occur at wall anchorages because of the stress concentrations that are created during ground motions (fig. 5.15). It is difficult to attach anchors to adobe walls successfully because the adobe itself is weak in shear and tension. To design more effective anchorages, it is important to understand the physical behavior of the adobe bricks and mortar materials around anchors.

Local section instability

Sections of an adobe building may become unstable after cracks have developed, and this is particularly true for sections of a wall that have become isolated from the building because wall openings are located too close to the corners. An example of this problem is shown in figure 5.16. The susceptibility to local section instability can be anticipated by evaluating the predicted general crack pattern that may result from an earthquake. The occurrence of cracks at openings and corners is

Figure 5.16
Local section instability: (a) illustration of how a local wall section may become unstable if the crack pattern results in an isolated block that can collapse; and (b) example of failure of garage wall section resulting from cracks at window and door openings (Andres Pico Adobe, Mission Hills, Calif.).

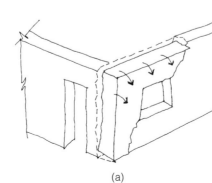

(a)   (b)

usually predictable, and the wall sections defined by the crack pattern can then be examined to determine which of them might become unstable during seismic ground motions.

Horizontal cracking in upper wall sections

Horizontal cracks may occur in upper wall sections if walls are anchored to the roof system or if a bond beam has been installed. Cracks can develop horizontally at or near the junction of the wall and the bond beam or roof framing as a result of either out-of-plane or in-plane movements, as illustrated in figure 5.17a. An example of this type of cracking is shown in figure 5.17b. A crack had developed at the bottom of a concrete bond beam, but the damage was not severe and the bond beam appears to have worked effectively. Another example of this type of failure occurred in the upper section of the walls of another historic adobe, shown in figure 5.17c. There appears to be some anchorage of the roof to the wall. Typically, with no attachment, the roof or ceiling framing can slip relative to the top of the wall before horizontal cracks develop.

Effect of Preexisting Conditions

Preexisting conditions may have a profound influence on the seismic performance of an adobe building. An assessment of the condition of any historic adobe building before an earthquake can help determine the types and potential extent of problems that may occur during a seismic event.

Figure 5.17
Horizontal cracking: (a) illustration of how lateral forces can result in a horizontal crack in the upper wall area when a roof or bond beam is attached to the wall; (b) example of horizontal cracks at the base of a concrete bond beam (Lopez Adobe, San Fernando, Calif.); and (c) horizontal cracking in upper wall section of the second story (Andres Pico Adobe, Mission Hills, Calif.).

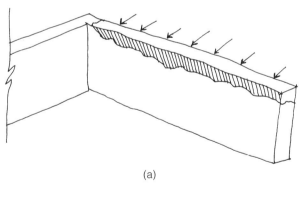

(a)

(b)

(c)

Moisture damage

Water is the most serious nonseismic threat to adobe buildings in areas of both high and low seismicity. It can damage an adobe wall by actually eroding away portions of the wall and by reducing the strength of the adobe material. *Basal erosion*, the disintegration and loss of a portion of an adobe wall at its base, can be caused by surface water runoff or by water falling from the roof and splashing up against the base of the wall. It can also be caused by water being drawn up into a wall by capillary action and then diffusing to the wall surface to evaporate. The water may contain soluble salts that crystallize near the surface as the water evaporates. In crystallizing, the salts expand and can fracture the adobe. Continuing deposition and crystallization of soluble salts slowly erodes the surface. The extent of basal erosion can be increased by the abrasive action of wind and sand, burrowing by insects or animals, and plant growth.

Regardless of the cause of the basal erosion, the result is that the area of the wall available to carry the loads imposed on it is reduced. When the loads exceed the compressive strength of the material, failure occurs. It is also conceivable that a wall could become sufficiently unstable to be subject to overturning if enough material is eroded from one face of the wall.

When the adobe at the base of the wall is weakened by moisture damage, a weak plane can develop, and the upper section of the wall can slip and collapse along this plane, as illustrated in figure 5.18a. This condition is most clearly shown in figure 5.18b, where a corner of a kitchen wall failed. The adobe at the base of the wall had been weakened by repeated exposure to moisture, which caused a weak failure plane to develop, and it appears that the wall slipped along this plane and collapsed. A similar failure mode was the probable cause of the failure shown in figure 5.18c. When a wall collapses, the location of the rubble can provide information on the probable location of the original failure. The wall shown in figure 5.18c appears to have collapsed down upon itself because the top of the wall is in the pile of rubble very near the original wall line. If overturning had occurred, the top of the wall would have been found some distance from the original wall location.

The major difference between the behavior of adobe and that of other masonry materials, such as brick or stone, is the dramatic reduction in strength when adobe becomes wet. Brick and stone can become saturated and still retain a large proportion of their strength, whereas long before adobe has reached saturation, its compressive and tensile strengths may have been reduced by 50% to 90%. This reduction in load-carrying ability can result in a material that can fail even under normal loads.

When moisture causes strength reduction to occur, adobe at first starts to deform slowly, and the rate increases as the adobe becomes wetter. A bulge at the base of an adobe wall is most often a sign of this settling or slumping. Repeated wet-dry cycles can also reduce the strength of the adobe significantly. When the clay component of the adobe repeatedly cycles from a moist to a dry state, the bonding between the clay particles and the other constituents of the adobe breaks down, which leads to a weakened material even after the adobe has dried.

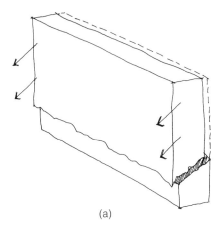

(a)

(b)

(c)

Figure 5.18

Illustration showing (a) how moisture damage can result in development of a plane of weakness along which a wall section can slide; examples showing how (b) moisture damage contributed to the collapse of the corner in this kitchen wall (Andres Pico Adobe), and (c) lower-wall moisture damage contributed to the catastrophic collapse of these two walls (Del Valle Adobe).

It is not necessary for an adobe wall to be wet at the time of an earthquake for water to have been a primary cause of failure. The lowered strength of water-damaged adobe results in a wall that is especially susceptible to damage or collapse. Spalling of adobe or cementitious stuccos can result from the combination of earthquake motion and a weakened bond between the adobe material and the surface rendering. This is shown in figure 5.19. If an entire wall section becomes wet or the adobe has been weakened by wet-dry cycles, the wall could fail suddenly (see fig. 5.18c).

Figure 5.19

Spallation of exterior stucco and adobe in water-damaged lower wall area (Andres Pico Adobe).

Out-of-plumb walls

Out-of-plumb walls can lead to wall overturning, which is probably the most serious effect that can occur to an adobe building during an earthquake in terms of life-safety hazard, the probable loss of historic fabric, and the generally high cost of building repairs. If an adobe wall is already out of plumb, it will be more susceptible to collapse than a nearly vertical wall. For example, if a wall is about 50 cm (20 in.) thick and about 2.5–5 cm (1–2 in.) out of plumb, and is not water damaged, it is not likely to be particularly sensitive to overturning. However, if the wall is 15 cm (6 in.) out of plumb, it may be exceedingly vulnerable to overturning during an earthquake unless it is firmly tied to other structural elements in the building.

Preexisting cracks

Preexisting cracks increase the susceptibility of a building to earthquake damage during moderate ground motions. These cracks may have been caused by previous earthquakes, wall slumping, or foundation settlement. An adobe building is likely to suffer extensive damage when the ground motion is intense (peak gravitational acceleration (PGA) ~ 0.4g), regardless of the condition of the building before the event. However, when ground motions are moderate (PGA ~ 0.2g), the extent of damage is heavily dependent on the condition of the building before the earthquake, and it is expected that the damage will be more extensive during moderate ground shaking if cracks are already present.

Chapter 6

Getty Seismic Adobe Project Results

The research that was carried out during the Getty Seismic Adobe Project was designed to provide an understanding of the seismic damage modes of adobe structures, to verify the appropriateness of taking a stability-based approach to retrofit designs for historic adobe buildings, and to further expand the knowledge of the details of implementing such retrofit systems. The research effort included a review of pertinent published and anecdotal information, analysis of historical and recent damage to adobe buildings, theoretical studies, and laboratory shaking-table testing of model adobe buildings. Because of the complexity of the dynamic behavior of adobe buildings, emphasis was placed on dynamic testing of models.

During the GSAP research period, the 1994 Northridge, California, earthquake occurred, and a survey of damage was conducted that provided invaluable information on the seismic performance of historic adobe buildings (Tolles et al. 1996). The combination of valuable field observations following the Northridge earthquake and the results from an extended dynamic research program made it possible to develop theory, tools, and techniques for retrofitting historic adobes that are both effective and respectful of historic fabric.

A brief review of the approaches taken in GSAP and the results obtained are given in this chapter and in appendix A. Complete details can be found in Tolles et al. 2000 and Ginell and Tolles 2000.

The Retrofit Measures Researched and Tested

During the GSAP testing program, the concept of stability-based design and some retrofitting measures that would confer these properties on adobe-walled buildings were evaluated. The principal retrofit measures investigated in GSAP were those that would have minimal impact on the historic fabric of the building. These measures included the following:

- upper- and lower-wall horizontal cables
- vertical straps
- vertical center-core rods
- partial wood diaphragms
- wood bond beams
- floor- and ceiling-level connections between walls, joists, and exterior horizontal cables

Horizontal and vertical straps or cables are designed to reduce the relative movement and displacement of cracked sections of the adobe walls during ground shaking. With such measures in place, adobe buildings could continue to remain stable while dissipating significant amounts of energy by friction. The stability provided would allow the structures to remain standing. Both straps and cables were connected through the walls using crossties.

Vertical, small-diameter, center-core rods were tested to evaluate a retrofit method that would not affect wall surfaces, which often have historically significant surface renderings. Test results showed that center-core rods were extremely effective in delaying the initiation of cracks, reducing the amount of irreparable damage, and generally improving the performance of a building. Although large-diameter (10–15 cm [4–6 in.]) center-core elements have been used in many other types of buildings, the diameter of the rods tested in this program were equivalent to 1.3–2 cm (0.5–0.8 in.) in full-scale buildings.

Partial wood diaphragms were used to test the hypothesis that full diaphragms are not required to provide stability to thick-walled structures, since little force is needed at the tops of the walls to prevent overturning. (For thin walls [$S_L > 8$], full diaphragms may still be required.) Anchorage to adobe is a serious problem because loads are concentrated at connections and can exceed the sheer strength of the relatively weak adobe material. For that reason, an alternative anchorage system was used that connected the attic-floor framing to a perimeter horizontal cable. During large ground motions, no signs of failure appeared around these ductile connections. Observations of failure at floor-level anchorages in recently installed retrofit systems were observed during the 1994 Northridge earthquake at the Pio Pico Mansion.

Some additional retrofit elements were tested briefly but were not developed fully. Saw cuts in walls were tested in an attempt to redirect the location of crack damage from a structurally critical site to one that was less important. Local ties were used to provide connections across cracks between existing cracked wall sections. However, placement of ties at potential crack locations requires knowledge of the building's specific crack patterns. Some crack locations are predictable, but others are random and depend on a combination of local construction details and stress distributions.

Anchorage systems at the tops of walls (e.g., dowels) similar to those used in conventional designs were also used in the tests but were not a focus of GSAP studies. The floor-level anchorage system using ties connected to the horizontal perimeter cable proved to be a very effective and ductile means of connecting the floor framing to the supporting exterior walls.

Dynamic shaking-table tests on model buildings were carried out to determine the effectiveness of selected retrofit measures and their influence on the overall stability of the models. A total of eleven model buildings were constructed and tested at a series of increasing levels of ground excitation:

- Models 1–6 were simple, four-wall, reduced-scale models (1:5 scale) without roof systems.
- Models 7–9 were small *tapanco*-type models (1:5 scale), which included attic, floor, and roof framing.
- Models 10 and 11 were two large-scale, *tapanco*-type models (1:2 scale). These were nearly identical in design to models 8 and 9. These models were instrumented to document the dynamic behavior of the buildings and to measure the stresses in selected elements of the structural retrofit system.

The primary purposes for using large-scale models were to gather numerical data on the buildings' dynamic behavior and to compare the performance of the large-scale models with that of the small-scale models, in particular, to evaluate the influence of gravity loading on failure modes.

Research Results Summary

The results of the research program clearly demonstrated the applicability of the basic theory that stability-based measures can be effective for both protecting the historic fabric and providing life safety in historic adobe buildings. The first few models tested demonstrated the general effectiveness of the retrofit measures, and the remainder of the research effort was directed toward parametric studies, identification of failure modes, and analysis of how these types of retrofit measures may work. Some of the numerical values obtained during the tests of the large-scale models can be used to estimate the maximum loads in similar retrofitting and structural elements. Elements can be designed using these values, but it is apparent that some engineering judgment must be applied in the case of real, incompletely characterized historic adobe structures.

In the design of a retrofit system, providing continuity throughout the structure is the most important aspect of the design, and in general the performance of the overall system is secondarily affected by wall thickness. Thicker-walled buildings ($S_L < 8$) are inherently stable when the adobe is undamaged and in good condition. Out-of-plane stability is due primarily to the resistance to rotation about the base of the walls. Only minimal restraint forces, which can be provided by horizontal straps or cables or vertical dowels at the tops of the walls, are required to provide out-of-plane stability.

Vertical straps or cables can be used to confer out-of-plane stability for thinner-walled adobe structures or to improve the ductility of these walls during continued ground shaking. Walls that collapsed in the unretrofitted models (models 8 and 10) did not collapse when retrofitted with vertical and horizontal straps (models 9 and 11), even when the walls were severely damaged.

Vertical center-core rods are extremely effective retrofit measures, especially when they are in continuous contact with the adobe walls. The effectiveness of this retrofit measure was particularly evident in the large-scale model (model 11), in which in-plane and out-of-plane damage was successfully limited to the formation of minor cracks.

Epoxy grouts proved to be effective in anchoring the rods to the adobe walls of the test buildings by virtue of the uneven penetration of the epoxy into the surrounding adobe. With this type of bonding, the center-core rods acted as reinforcements and actually strengthened the wall assemblies. Even if the bonding between the adobe and the epoxy is not particularly strong, center-core rods function well because they act as effective shear doweling between adjacent cracked wall sections. In doing so, center-core rods greatly improve both the in-plane and out-of-plane strength and stability of adobe walls.

Chapter 7

The Design Process

Before plunging into the retrofit design process, the design team must devote some effort to identifying the goals that can be attained by retrofitting. A wide range of possible goals and combinations of options exists, and as a first step a rational selection should be made based on priority judgments. All building codes specify that the highest priority of a seismic retrofit design is to provide for life safety. For historic adobe buildings, additional design goal options are to provide a retrofit that

- has a minimal effect on the integrity of the historic fabric;
- minimizes change to the building's appearance;
- can be removed with minimal effect on the structure or its appearance;
- selectively protects specific architectural or historic features;
- directs damage toward areas of lesser significance;
- minimizes damage during minor or moderate earthquakes (Richter magnitude < 6); and
- minimizes structural damage during major earthquakes (Richter magnitude > 6).

Although all these options are desirable design goals, choices must inevitably be made. Goals need to be selected that (1) are compatible from an engineering standpoint, (2) can be accomplished within a given budget, and (3) are consistent with the priorities established by the planners. Selection of the principal design goals will determine the type of intervention or combination of interventions that would be most suited for consideration at a particular site. The following are some of the possibilities:

- **Minimum levels of intervention:** For structures that have moderately thick or thick walls, it may be possible to attain reasonable levels of seismic safety and significantly reduce the life-safety hazard simply by using anchors at the tops of walls.
- **Moderate levels of security and intervention:** A more detailed design may include vertical and horizontal straps, structural redundancy, strengthening of the roof system and/or addition of bond beams.

- **High levels of security and damage control:** Center-core rods coupled with other retrofit measures can be used to greatly increase levels of safety, reduce the potential hazard to the historic fabric, and reduce the likelihood of severe structural damage.

This is not an exhaustive list, but it provides some idea of the range of solutions that may be offered by the retrofit designer.

The design process should follow a logical sequence:

1. Develop a global design strategy that will provide continuity to the building, and design a system that ties the building together. It is important that the basis of the retrofit design be one of allowing the building to function as an integrated system.
2. Predict the location and pattern of cracks that may occur during major earthquakes, and address the potential stability and probable failure modes of each major cracked wall section.
3. Design retrofit measures that can assure the stabilization of each cracked wall section and limit permanent damage to acceptable levels.

These basic design issues are the primary subject of this chapter. First, however, it is important to consider the subject of earthquake severity.

Designing for Earthquake Severity

When beginning to consider retrofit designs for an occupied adobe building, one should first obtain information on the probability of earthquake occurrence, the characteristics of the high- or low-magnitude earthquakes that can be expected to occur, and the types and location of historic building fabric elements that require maximum protection. These factors should be considered when formulating a global design that provides overall structural continuity (see "Global Design Issues," next section) and minimizes the occurrence of the most prominent failure modes—such as wall overturning, mid-height wall collapse, corner out-thrusting or collapse—or failures that arise from a combination of in-plane and out-of-plane motions. Adobe buildings can be rather resilient structures during earthquakes if the walls and roof of the building are simply tied together.

Life-safety concerns should be addressed regardless of whether the projected design-level earthquake is major or minor. However, the analysis performed during the design of a seismic retrofit system should provide for life safety during the largest predicted seismic event.

For minor to moderate, more frequent earthquakes

A retrofit strategy may be chosen that would minimize the extent of damage. Although a stability-based retrofit system will greatly enhance life safety, it may do little to minimize damage during these earthquakes. Damage during minor and moderate earthquakes may be affected by measures that have little effect on the overall structural stability. For example, in a building containing preexisting earthquake damage cracks and in which a stability-based retrofit system has been implemented, it may be desirable to pressure-grout and repair the cracks. Whereas filling the existing cracks may have little effect on structural stability during larger seismic events, it may have a profound effect on building damage during more moderate earthquakes.

For major earthquakes

It may be difficult to reduce the overall extent of damage during severe earthquakes, but some retrofit measures may provide greater structural damage control than others. These measures can be more expensive and invasive than others, but their incorporation into the design of the retrofit may reduce the severity of damage during major events and can prevent catastrophic losses.

Global Design Issues

Recognition of global design issues is the starting point in the design process. The basic elements of global design are

- upper-wall horizontal elements (mandatory)
- vertical wall elements (optional except for thin-walled structures)
- lower-wall horizontal elements (optional)

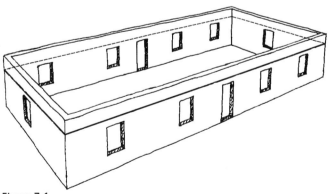

Figure 7.1
Illustration showing upper-wall horizontal elements used to prevent out-of-plane overturning and in-plane offsets. These can be a bond beam, straps in conjunction with the floor or roof system, or a partial diaphragm.

Upper-wall horizontal elements

These elements are the most important part of a seismic retrofit for an adobe building because the principal mode of wall failure is overturning; upper elements are designed to prevent this type of failure. Therefore, the initial step in the global design is to provide upper-wall horizontal elements, as shown conceptually in figure 7.1, that can perform the following functions:

- provide anchorage to the roof or floor
- provide out-of-plane strength and stiffness
- establish in-plane continuity

Three possible types of upper-wall elements are (1) partial plywood diaphragm, (2) concrete or wood bond beam, (3) external nylon or steel straps or cables combined with existing, flexible roof or floor framing. Installation of concrete bond beams often involves removal of a substantial amount of original roof framing and a potentially large loss of historic fabric; the loss would not be as great if wood bond beams were installed.

In addition to preventing overturning, the upper-wall elements should also provide in-plane continuity, which prevents cracked wall sections from moving apart in the plane of the wall. Either a bond beam or horizontal straps or cables will provide in-plane continuity because they are continuous elements along the length of the wall. A partial plywood diaphragm may consist merely of a 4-foot width of plywood nailed along the tops of the joists. Chord members should be provided with partial plywood diaphragms similar to those used in standard diaphragm design. Chord members provide in-plane continuity along the length of the wall and can act as carriers for the "flange" forces of the partial plywood diaphragm beam. Vertical dowels, grouted into the wall and connected to a bond beam or roof diaphragm, are highly effective in promoting continuity.

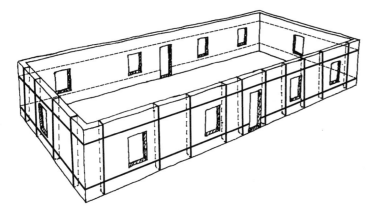

Figure 7.2
Diagram showing how vertical elements add resilience and redundancy to the structural system and restrict the displacement of cracked wall sections. These can be surface straps or internal center-core rods.

Vertical wall elements

Vertical wall elements can greatly improve the resilience of a structure during extended ground motions and can help minimize the extent of damage during major earthquakes. The thickness of the walls and the level of security desired will determine whether vertical elements should be used. Vertical wall elements, as exemplified by nylon straps, steel straps, or steel cables, should be attached to both interior and exterior wall surfaces. The use of vertical elements can greatly increase the "ductility" of the walls, as shown in combination with upper and lower wall elements in figure 7.2.

The need for vertical wall elements is most important for thinner adobe walls. Thick walls are unlikely to need vertical straps, although center-core rods may help reduce shear displacements. Moderately thick walls can be retrofitted with vertical elements to improve wall performance and minimize offsets during extended seismic events.

Thin adobe walls ($S_L \geq 8$) require vertical straps (fig. 7.3), center-core rods (fig. 7.4), or some other type of treatment to prevent out-of-plane failure. Center-core rods, grouted into oversized holes using an epoxy, polyester, or cementitious grout, or vertical straps on both sides of a wall, can be used as reinforcing elements to prevent out-of-plane failure. Grouted center-core rods bond well to adobe because the grout material is absorbed irregularly into the adobe. For thicker adobe walls, center-core elements tend to act as shear dowels rather than as flexural reinforcements. Although the wall behavior when dowels are used could be analyzed in terms of a flexible reinforcement, the principal function of the center-core rods is that of shear dowels, and an analysis based on this function should be used.

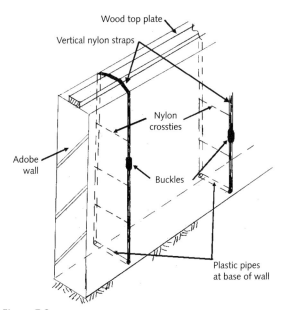

Figure 7.3
Diagram of vertical straps and crossties on adobe wall.

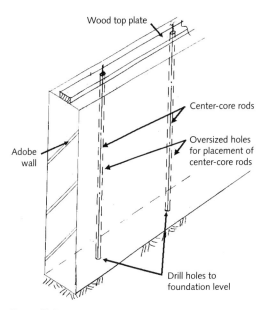

Figure 7.4
Diagram of center-core rods in adobe wall.

The diameter of center-core rods used in full-scale walls can range from 12 to 25 mm (0.5–1 in.), and the rods should be inserted in holes that are sized according to the needs of the material used for anchoring the center-core rods. The GSAP research was based on a prototype adobe wall that was 41 cm (16 in.) thick, and 17 mm (0.67 in.) diameter center-core rods worked well even when used without grout. Small-diameter rods and holes less than 50 mm (2 in.) in diameter should be used because larger-diameter center-core elements may act as "hard spots" and serve to split the low-strength adobe wall.

Lower-wall horizontal elements

Lower-wall horizontal elements can be used to improve the performance of adobe walls by preventing cracked wall sections from "kicking out" in plane, along the length of the wall. In some instances, a wall section may be displaced into a door opening, but more serious problems tend to occur at the ends of the walls where cracked wall sections are unrestrained and can move outward at the base. In some instances, repairs may consist only of filling cracks, but in other cases, if it is not adequately restrained, the entire wall may need to be reconstructed. A schematic design using upper- and lower-wall horizontal elements is shown in figure 7.5.

Lower-wall horizontal elements can consist of straps or cable elements or even buttresses. One of the critical features of lower-wall horizontal elements is the end connection, because large loads can be imposed when wall sections tend to move outward. An effective means for providing this type of support is the use of center-core elements for stabilization of wall sections that may fail along diagonal cracks, as shown in

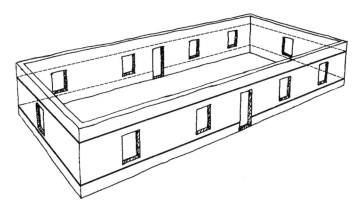

Figure 7.5
Illustration showing how lower horizontal elements prevent displacements at the base of the walls.

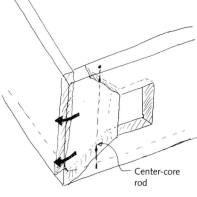

Figure 7.6
Diagram showing stabilization of corner
wall section near a wall opening by use
of a center-core rod.

figure 7.6. The core element extends across the diagonal crack and prevents slippage of the cracked wall section by acting as a shear dowel.

Crack Prediction

During earthquakes, adobe walls crack into large sections. Most collapses are localized and occur because of the instability of the cracked
wall sections. Each section can be displaced and overturn independently
of the remainder of the building. However, if the basic crack patterns of
a structure are predicted first, the overall retrofit system can be designed
to stabilize the numerous wall sections that may collapse or sustain substantial permanent damage. After cracks have developed, the behavior of
the building is largely dependent upon the stability of the cracked wall
sections, and the design of the retrofit system should be directed toward
stabilization of each of the sections. Wall sections are formed by cracks
that develop at wall openings, at corners and other wall intersections, at
mid-walls, and at locations such as regions of material incompatibilities.
A predicted crack pattern is shown in figure 7.7 (see chap. 5 for a discussion of typical damage typologies).

Although the location of major cracks often can be predicted,
other potential cracking areas are difficult to identify. Some cracks may
already exist due to water intrusion, foundation settlement, or wall
slumping or may occur in unforeseen areas.
The inclusion of additional vertical elements
may be advisable to provide for redundancy
in the structural system.

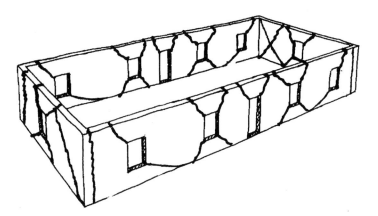

Figure 7.7
A map of potential adobe-block formation, resulting from crack pattern prediction, a necessary prerequisite to the
design of a stabilizing retrofit system.

Retrofit Measures

The selection of retrofit measures for a specific
structure requires the consideration of the
expected orientation of earthquake-induced
forces with respect to details of the building's
construction.

Out-of-plane design
The design of the retrofit system for controlling out-of-plane wall displacements is the single most important aspect of the retrofit design
because out-of-plane collapse (overturning) is a costly, catastrophic, and
life-threatening type of failure. Thicker adobe walls are more resistant to
overturning than thinner walls, and minimal forces are required to stabilize the tops of thick and moderately thick walls. Therefore, attachment
of a wall to an upper-wall element or roof system is critical to the design;
in this case, the need for strengthening or stiffening with a diaphragm is
not an important design consideration.

Elastic analysis methods can be used to understand the interaction between an adobe wall and upper-wall horizontal elements.
However, if an elastic analysis were used to design a retrofit that would

transfer the dynamic forces from the out-of-plane to the in-plane walls, the resulting horizontal element would be extremely rigid and likely to cause additional problems. A stiff upper-wall element (such as a strong bond beam) may transfer a very large portion of the lateral load into the transverse walls, thus overloading these walls and causing in-plane shear damage that would be difficult to repair.

In full-scale (prototype) dimensions, the wood bond beam used in some of the GSAP small-scale shaking-table tests was only 5×19 cm (2×7.5 in.) and spanned a distance of 7 m (23 ft.) horizontally. In the elastic region, this bond beam would have a negligible effect on the dynamic behavior of the walls. After cracks had developed, the presence of a bond beam would have a large impact on the out-of-plane behavior of the walls. The strength and the stiffness of the wood bond beam actually used in the tests were more than sufficient to transfer loads to the in-plane walls.

It was shown that only small forces were required to prevent the out-of-plane displacement of moderate to thick adobe walls (Tolles and Krawinkler 1990). Simply anchoring the tops of the walls to the roof rafters prevented collapse of out-of-plane walls, and thus the horizontal stiffness that could be imparted by a diaphragm was not needed. The addition of only minor additional horizontal stiffness would have prevented the out-of-plane failure of these walls that occurred at high levels of acceleration and larger displacements. Therefore, some limited out-of-plane stiffness is required to prevent the overturning of moderately thick adobe walls ($S_L = 6$–8). Thin adobe walls ($S_L > 8$) are not resistant to overturning and should be fitted with vertical-wall reinforcing elements, such as center-core rods, with full diaphragm support at the tops of the walls. For thin walls, the bond beam or roof diaphragm will be the most important factor in determining the dynamic behavior and, ultimately, the stability of the building.

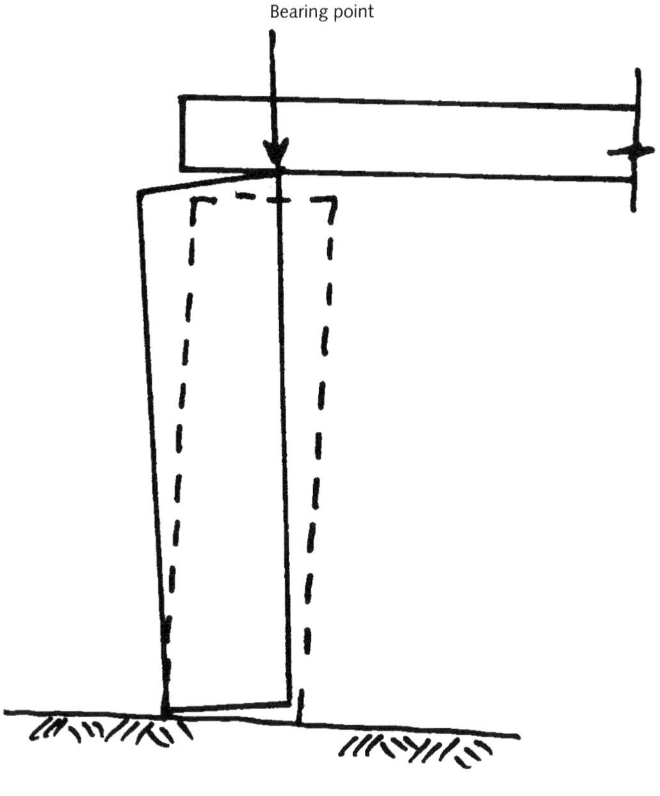

Load-bearing and non-load-bearing walls

Load-bearing adobe walls are more resistant to damage in earthquakes than are non-load-bearing walls. The improved performance is the result of the stabilizing effects provided by the framing that is supported by the walls. This occurs even when a positive connection between the framing and the walls does not exist, which is typical of historic adobe buildings constructed during the 1800–1900 period. Framing will result in additional lateral restraint for two reasons: it provides direct resistance to the out-of-plane motions of the walls, and framing exerts a downward force on the wall as the wall starts to rock upward (fig. 7.8). The vertical force resulting from the weight of the roof system provides stability to the wall by resisting the overturning forces.

Figure 7.8
Illustration of thick, load-bearing walls, which are more stable than non-load-bearing walls because of the upper restraints provided by roof or floor framing.

Bearing point

Gable-end walls

Gable-end walls are the walls of an adobe building most susceptible to collapse. First, these walls are taller than others in the building, but are usually of the same thickness. Second, gable walls are typically non-load-bearing, and the roof, attic, and/or floor framing provides little restraint against outward motion. Gable-end walls should be securely anchored to the building at the roof and the attic floor levels for out-of-plane stability. Center-core rods are especially useful for preventing out-of-plane collapse of these walls.

In-plane design

The most important design feature that can improve the postelastic, in-plane performance of a wall is to provide for in-plane continuity along its length. Large amounts of energy can be dissipated within in-plane walls if the cracked wall sections are held together by continuity elements that allow the sections to move and dissipate energy through friction without allowing the wall to deteriorate. An anchored bond beam can add substantial continuity along the top of a wall, and upper or lower horizontal wall straps can hold the cracked sections of wall together and prevent extensive wall degradation. The ultimate capacity of shear walls can be greatly improved by the addition of in-plane continuity elements. It was shown experimentally (Tolles and Krawinkler 1990) that a wall with in-plane continuity outperformed a wall with twice the "shear area" that lacked this feature.

Significant improvements in the performance of shear walls can be provided by center-core rods, which act to increase the damage threshold level, reduce the amount of cracking during extended shaking, and increase the ductility of the structural element. Given this improvement, the static design analysis could include an increase in allowable shear for adobe walls reinforced with center-core rods.

The failure patterns observed in the model buildings tested in GSAP were also observed in historic adobe buildings after the 1994 Northridge earthquake (Tolles et al. 1996). The most common type of serious damage occurred at diagonal cracks at the corner of a building (see fig. 7.6). The cracked wall section adjacent to the corner can slide both outward and downward. This type of crack may be difficult and costly to repair and may lead to instability of the entire wall if the offset is large enough.

A preferable alternative to a simple static design analysis is to allow thick, out-of-plane walls to rock but to include some restraint by the in-plane walls. However, forcing large loads from out-of-plane walls into in-plane walls may easily overload the in-plane walls and result in another type of damage that is difficult to repair.

After the Northridge earthquake, numerous adobe walls were observed to have the classic crack patterns characteristic of out-of-plane rocking. Walls that are 61 cm (24 in.) thick can easily rock 15 cm (6 in.) or more in either direction, but they will tend to come back to rest in their original position. Only nominal resistance is required to prevent thicker adobe walls from overturning. The design of a diaphragm system will be dependent upon the type of analysis used to determine the distribution of loads between the in-plane and out-of-plane walls for each direction of loading.

The in-plane performance of walls with center-core rods was significantly better than that of any other form of retrofit. The improved performance was particularly significant in the large-scale GSAP model tests, where cracks were observed to terminate at center-core rod locations. In-plane walls retrofitted with center-core rods sustained very little damage, and no offsets occurred even during the largest table-displacement dynamic tests.

Diaphragm design

The design of a diaphragm system for historic adobe buildings is different from that used for most other buildings. In conventional design, the purpose of a diaphragm is to transfer loads for a given direction of motion from the roof and out-of-plane walls to the in-plane walls. The walls of conventional buildings are weak out of plane and strong in plane, and therefore diaphragms are required that can transfer loads.

Designs for thicker-walled historic adobe buildings should be different. Such walls have substantial stability in the out-of-plane direction, and the forces required to stabilize the walls should be relatively small. As long as an adobe wall remains erect, the damage during out-of-plane rocking is not severe. Typically, damage that does occur is confined to horizontal cracks along the base and to vertical cracks adjacent to perpendicular walls. Stability is directly related to the absolute thickness and the height-to-thickness ratio of the wall.

The second factor affecting diaphragm design for historic adobe buildings is the in-plane capacity of the adobe walls. If a diaphragm is stiff, loads will be transferred from the out-of-plane walls to the in-plane walls for a given direction of motion. As a result, the in-plane walls can be overloaded and severely damaged. Although the out-of-plane walls can easily rock 15 cm (6 in.) back and forth, a displacement of 1.25–2.5 cm (1–2 in.) in the in-plane walls can produce significant and costly damage. Therefore, the diaphragm stiffness, which determines its load-transferring ability, need not be great, but it should be sufficient to provide stabilization for the out-of-plane walls.

Again, thin-walled adobe buildings do not have the same resistance to overturning as that of thick walls. These buildings should incorporate diaphragm systems similar to those used in more conventional types of construction.

Connection details

Connection details in adobe walls in particular must be properly designed, because adobe is such a low-strength material. Connections can fail at their bearing locations and thus not perform as intended. Stress concentrations occur at anchorage locations and result in cracking at or near these sites. However, cracks themselves are often not particularly significant, unless they lead to other damage or instability.

The fundamental guidelines involved in connection design for adobe walls are these:

- Distribute the load whenever possible to decrease stress concentrations.
- Predict potential crack patterns, assess the impact of these cracks on stability, and provide redundant retrofits as required.
- Avoid brittle connections; design the details to allow for ductile behavior.
- Include redundancy wherever possible.

This section presents a number of suggestions for details in the retrofit of historic adobe structures. Photographic illustrations show the types of damage that may occur, and drawings and details are included to illustrate what *not* to do. Other details are shown to demonstrate how to construct resilient and ductile connections in adobe walls.

Wall anchorage

Anchorages in adobe walls can result in stress concentrations that could lead to local adobe failure. Adobe resists distributed loads more easily than point loads. Energy dissipation can be a useful concept, and some anchorage methods can be used to cause localized damage in the adobe without causing cracking of the wall. If designed properly, this type of connection will dissipate much more energy than a brittle connection. Multiple connections or a second type of connection can provide redundancy. Using multiple connections distributes the load, and adding a second type provides a backup for the first.

Knowledge of failure modes can help identify the location for a secondary connection or for a member that can provide resistance in case failure occurs. If a limited level of damage around connections can be tolerated, other backup elements may not be needed. On the other hand, if failure of a connection will lead to instability, then a secondary, redundant connection or detail is highly recommended.

Roof-to-wall connections

Roof framing in historic adobe buildings is often only lightly attached to the walls. Very often, the framing is not actually attached at all and just rests on top of the wall, as shown in figure 7.9. Thus, the roof framing can slide relative to the wall or can dislodge bricks at the top of the wall.

Figure 7.9
Roof framing resting on, but not attached to, the top course of adobe bricks.

Attaching the roof framing to the wall is important for achieving overall structural stability and for preventing relative roof-wall movement. This is generally accomplished by means of steel dowels.

The problem with top-of-wall anchors is that, while they are critical to the structural performance, they tend to result in brittle connections. Multiple anchors between the roof system and the wall can help to distribute the load more evenly along the tops of walls. An upper-wall horizontal cabling system can serve as a redundant element in case the top-of-wall anchors fail.

Floor-to-wall connections

Floor-to-wall connections can be difficult to implement but, when done properly, can be effective because of the significant overbearing pressure from the wall above. Where no anchorage exists, slippage of the floor joists relative to the wall may occur (see chap. 5, fig. 5.14). One example of an effective floor-to-wall connection is shown in the details at the second-floor level of the Andres Pico Adobe. This connection was effective during the 1994 Northridge earthquake, despite heavy damage to much of the building. An exterior continuous plate was attached to the through-wall floor joists. Lag screws were anchored into the end grain of the joist (fig. 7.10) to prevent the relative movement of the walls and the floor joists. Bolting into the end grain of a joist is not generally recommended, but in this case, the joint had sufficient strength to prevent relative movement during the earthquake. This detail is diagrammed in figure 7.10b.

Attachment of the floor joists to a perimeter horizontal cable is another effective means of anchoring the floor joists to the walls. In the GSAP tests, straps were used to tie the floor joists, through the exterior wall, to the perimeter horizontal strap (fig. 7.11). The connection was not rigid, but it prevented serious damage to the structure. A similar connection was used at the Del Valle Adobe at Rancho Camulos, and in this case, the floor joists were attached to eyebolts through which an

Figure 7.10
Floor-to-wall connections: (a) side of wall, showing continuous ledger and lag screws driven into the end of a joist; and (b) section diagram showing lag screw, ledger, and joist.

(a)

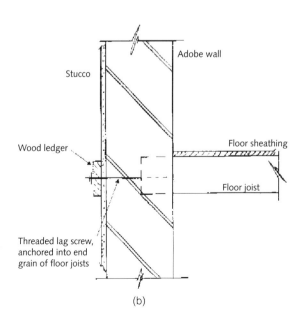

(b)

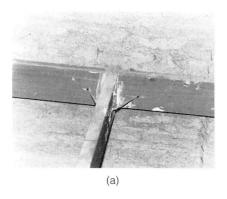

(a)

12 Slope
◁ 8

Partial plywood diaphragm

Sheetrock screws used as
anchor bolts extended at least
three courses into the adobe wall

Lag screws between roof
rafter and discontinuous plate

Blocking

Partial plywood
diaphragm

Roof rafters, ⅝″ × 1″

Floor joists, ¾″ × 2″

Discontinuous plate anchored
to wall with Sheetrock screws
and screwed to roof rafters

Adobe wall

Exterior strap

Through-ties

(b)

Figure 7.11
Connection systems: (a) attachment of
floor joist to exterior horizontal cable can
provide an effective connection (GSAP
model 11); and (b) cross section of wall-
roof-floor retrofit that was tested on a
shaking table.

exterior perimeter cable was placed (fig. 7.12). This detail is shown in
figure 7.13a. Another option, shown in figure 7.13b, can be used if the
floor joists are not visible from below or if visibility of the attachment is
not a significant factor.

Another type of anchoring system, used at the attic-floor level
of the Pio Pico Mansion, failed during the Northridge earthquake, where
the relatively minor peak ground horizontal acceleration was less than
0.15g. A section detail of this anchorage is shown in figure 7.14a. A flat
strap was anchored into the wall using an adobe-mud–fly-ash mixture.
The anchorage seemed to have worked, except that the entire plug pulled
out of the wall, as shown in figure 7.14b. Loads on these connections to
the gable-end wall can be very large. This type of connection was brittle
and had no redundant elements in other parts of the retrofit system to
serve as backup.

Figure 7.12
Retrofit at Del Valle Adobe, Rancho
Camulos, Piru, Calif.: (a) interior view
of eyebolts used to attach the exterior
horizontal cable to the floor joists, and
(b) exterior view of the eyebolts.

(a)

(b)

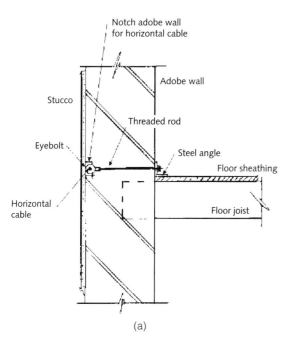

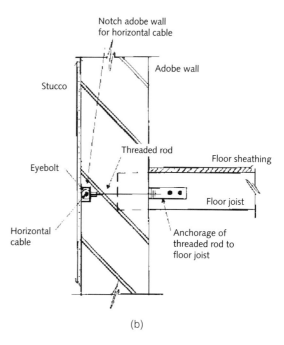

(a)   (b)

Figure 7.13
Diagrams of Del Valle Adobe retrofit:
(a) cross section of eyebolt attachment between the horizontal cable and floor sheathing and joist, and (b) cross section of eyebolt attachment between the horizontal cable and joist—alternative method.

Figure 7.14
Pio Pico Mansion, Whittier, Calif.: (a) section diagram showing steel flat-strap connection anchored into an adobe gable-end wall using a low-shrinkage mud–fly ash grout, and (b) photo taken after the 1994 Northridge earthquake; the connection failed when the grout plug pulled out of the wall.

Wall-to-wall connections

Connections to enhance lateral support for adobe walls are common. Nevertheless, if connections are stiff, the forces generated locally can be very large, and as a result, the connections can fail or not function as intended or unforeseen damage to a wall may occur.

Tie-rods are often installed in historic adobe buildings either after earthquake damage or beforehand, to forestall overturning of thin, parallel walls. Tie-rods are generally threaded steel rods that are secured by nuts and stress-distribution plates on the exteriors of parallel walls. Tie-rods can add needed support, but may also create extended damage. Figure 7.15 shows the numerous repairs that were carried out at the end of a tie-rod in an attempt to repair damage. The anchorage shown in figure 7.16a has pulled into the wall and has allowed the steel tie-rod to sag and lose its effectiveness, as shown in figure 7.16b. Anchorages can also lead to wall

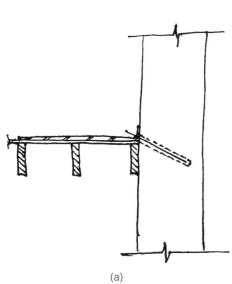

(a)

(b)

(a)   (b)

Figure 7.15
Numerous repairs to wall at tie-rod end plate (De la Ossa Adobe, Encino, Calif.).

Figure 7.16
Photos showing (a) tie-rod stress-distribution plate pulled into the wall; and (b) sagging, ineffective tie-rod.

Figure 7.17
Damage to wall adjacent to wood anchor (Rancho San Andres Castro, Watsonville, Calif.).

damage adjacent to the anchor. The severe cracking at the anchorage shown in figure 7.17 is typical and has led to the instability of the wall to the left of the anchorage. This type of tie has often been used in an attempt to stabilize the out-of-plane walls of unreinforced masonry buildings. Unfortunately, these ties can encourage vertical cracking of end walls that can enhance the instability of such non-load-bearing walls.

Achieving effective connections between perpendicular walls can be difficult because of the very different in-plane and out-of-plane motions of thick adobe walls. Connections between intersecting walls can have sufficient strength to withstand moderate ground motions if the original construction consisted of overlapping bricks or contained overlapping reinforcements. Regrouting or doweling of existing cracks gives the structure some ability to withstand moderate ground motions. An illustration of this type of repair is shown in figure 7.18a. However, during strong ground motions, damage will occur to adobe walls at or near the junction. Although regrouting and doweling may result in a significant benefit during moderate ground motions, they will have little effect during very large ground motions. The cracks that would have occurred at the corners, if doweling had not been present, would probably occur just beyond the location of the dowels, as illustrated in figure 7.18b.

Although local anchorages between walls can be an effective means of limiting damage, they are likely to fail during major earthquakes, and other elements of the global retrofit system should be designed to become active after these cracks have developed. The horizontal elements of a global retrofit system may be exceedingly effective in providing structural stability once cracks have developed at wall intersections. Local anchorages could be combined with a global strapping solution to provide a complete set of measures that would be effective during moderate and major seismic events.

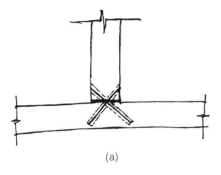

(a)

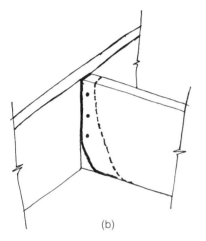

(b)

Figure 7.18
Illustrations of regrouting and doweling, showing how (a) anchoring dowels can be effective at perpendicular interior walls during moderate ground motions; and (b) during strong motions, cracks can occur adjacent to the dowels (dotted line).

Redundancy

The deliberate addition of alternate paths for the distribution of earthquake-induced forces is known as *redundancy*. If only one path exists, catastrophic results can occur if that path is lost. Because the seismic engineering analysis of adobe structures cannot yet be considered to be a precise science, the a priori prediction of the seismic behavior of an adobe building or the location of cracks in a given structure is still somewhat limited. However, inclusion in the design of alternate, redundant paths for the transfer of forces could provide greater confidence in the ability of the retrofit design to minimize serious damage in unanticipated areas.

Moisture problems

Moisture damage at the base of adobe walls is a common problem that must be addressed and solved as a part of any retrofitting design. Some results of moisture-related problems are shown in figures 7.19 and 7.20. Sliding of an entire wall section (see fig. 5.18a) is another possible consequence of water damage. The resistance of thick adobe walls to overturning is greatly diminished when moisture damage is not corrected. Deteriorated adobe bricks at the base of walls (see fig. 5.19) may need to be replaced before the inherent resistance to overturning provided by the thickness of the walls can be realized.

Figure 7.19
Example of adobe wall slumping and washout resulting from moisture damage (photo courtesy Tony Crosby).

Figure 7.20
Example of wall collapse due to water damage at base (photo courtesy Tony Crosby).

Chapter 8
Design Implementation and Retrofit Tools

Although the emphasis of the GSAP research was directed toward evaluation of stability-based retrofit designs, strength-based designs that involve lateral analyses also should be part of a complete design documentation. A brief discussion of standard lateral design procedures as they apply to adobe walls is given in the next section. This is followed by some specific elements of wall design and a discussion of the implementation of several of the retrofit measures studied during the GSAP research effort. A few simple design examples showing the relative effects of wall thickness are provided.

Standard Lateral Design Recommendations

The GSAP research did not specifically address issues associated with the basic design force levels as prescribed by building codes such as the Uniform Building Code (UBC), the Uniform Code for Building Conservation (UCBC), or the California State Historic Building Code (SHBC). Nevertheless, all design procedures should typically include static or dynamic analyses for determining forces that are distributed to and resisted by the shear walls.

The level of seismicity to be expected in the area where a building is located defines lateral force levels. The seismic map of the United States shown in figure 8.1 was taken from the UBC (1997). Lateral forces, as prescribed by the 1997 UBC, may be higher than those given in previous versions of the UBC because the more recent edition provides for near-source effects (Structural Engineers Association of California 1999). The codes define earthquake faults of Types A and B and zones that are 5, 10, and 15 km from the faults (fig. 8.2). If the site lies within these zones, the seismic loading factor is increased accordingly.

A principal conclusion of the GSAP study and a fundamental tenet of the design methodology is that elastic design procedures often do not predict the ultimate behavior of unreinforced masonry buildings. Therefore, strength-based design procedures should be used cautiously, and understanding how an adobe building might collapse and how to prevent the collapse is much more important than determining the exact stress level in the shear walls.

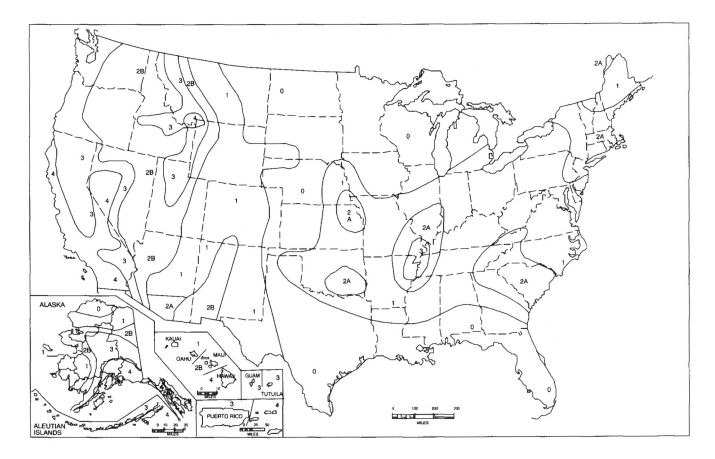

Figure 8.1
Seismic zone map of the United States of America (UBC 1997).

Nevertheless, some observations that were derived from the test results should be made regarding design-level shear stresses in adobe shear walls. Two simplified procedures, based upon the UCBC and SHBC, were used for determining the shear stresses in the adobe walls. Where upper-wall continuity elements were used, the in-plane adobe walls performed satisfactorily during even the largest dynamic tests involving both large-scale and small-scale model buildings. Severe cracking occurred in many buildings, but the in-plane shear walls did not fail.

The SHBC allows shear stresses of 28 KPa (4 psi) in adobe walls and uses UBC design force levels for static design. In Zone 4, which includes a large part of California, these force levels would be between 14% and 18% of gravity (0.14–0.18g). The UCBC allows a lower design force level of 10% (0.1g) for buildings with an occupant load of fewer than one hundred persons and 13.3% (0.133g) for buildings with a greater occupant load, but then the allowable design shear stress is only 21 KPa (3 psi). When applied to the model buildings, both design procedures are roughly equivalent. The SHBC allows higher stresses, but the design forces are also higher.

The design stresses in the shear walls of the model buildings were checked using each design procedure and were found to be just less than those prescribed by the SHBC and UCBC. Since all in-plane shear walls of the model buildings survived the strongest dynamic tests, the design procedure used is considered to be adequate for general design, although the in-plane shear walls suffered significant shear damage in the large-scale model tests. From this limited information, it can be stated

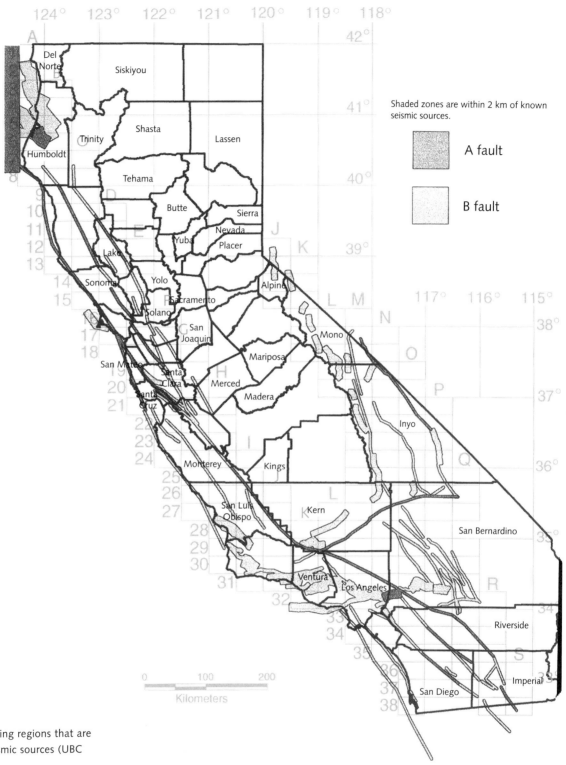

Figure 8.2
Map of Calif. showing regions that are close to known seismic sources (UBC 1997).

that these walls were not greatly overdesigned or underdesigned. Therefore, any significant change in the allowable shear-stress or force design levels (without substantial retrofit measures that increase ductility) is not recommended.

Wall Design

Out-of-plane wall design

As discussed earlier, the design of the retrofit for an adobe wall is greatly affected by the wall thickness, and thin walls require much higher levels of intervention than thick walls.

Thick walls, however, are as susceptible to shear cracks as they are to out-of-plane cracks. The principal retrofit effort should focus on a system that ties the structure together. A thick wall ($S_L < 6$) can still overturn, and so anchorage of the tops of these walls to the roof system is required. The following retrofit recommendations for thick walls may be made, based on the design level of safety and the potential for permanent structural damage:

- **Low and mid safety levels:** Anchor the walls to the roof system without further reinforcement of the walls themselves.
- **High safety and minimal damage levels:** Anchor the walls to the roof system, and reinforce the walls with center-core rods either at the corners or along the length of the walls. Center-core rods will minimize the extent of shear offsets that may occur either in plane or out of plane.

Moderately thick walls ($S_L = 6–8$) are susceptible to shear cracks and are unlikely to suffer mid-height, out-of-plane failure. Out-of-plane cracks are likely to occur before in-plane cracks develop. Since there is little chance of mid-height, out-of-plane failure, no reinforcement is required for these walls at minimum safety levels. The following retrofit recommendations may be made based on the level of safety and the potential for permanent structural damage:

- **Minimal safety levels:** Anchorage of the walls to the roof system to prevent overturning. No additional wall reinforcement required.
- **Moderate levels of safety:** Anchor walls to roof system and use vertical straps at regular intervals; this greatly increases the ductility of the structural system and reduces the chances of progressive wall failure.
- **High safety and minimal damage levels:** Anchor the walls to the roof system and reinforce walls with center-core rods at regular intervals. Center-core rods will minimize the extent of shear offsets that may occur either in plane or out of plane.

Thin walls ($S_L > 8$) are inherently unstable and can fail by rotation about the base. They may also collapse due to mid-height, out-of-plane failure that may occur before in-plane cracks develop. Therefore, vertical reinforcing elements are required for thin walls. The following retrofit recommendations may be made based upon the level of safety required and the potential for permanent structural damage. Thin, out-of-plane walls *must have* some type of retrofitting.

- **Minimal to moderate safety levels:** Anchor walls to roof system and use vertical straps at regular intervals to ensure that the building does not collapse during strong earthquake motions. Retrofitted walls may degrade substantially during long, sustained, strong earthquake motions, but failure is unlikely.
- **High safety and minimal damage levels:** Anchor the walls to the roof system and use center-core rods at regular intervals. This will allow walls to perform well both in plane and out of plane while also minimizing shear offsets that may occur either in or out of plane.

In-plane wall design

Calculated in-plane shear stresses are often a controlling factor in the strength design of adobe buildings. Since the out-of-plane motion of moderate to thick adobe walls is largely resisted by rotation about the base, the calculated values will be higher than actual forces from out-of-plane walls. If the calculated in-plane shear stresses are higher than acceptable values, the use of center-core elements within existing adobe walls may justify an increase in design stresses for adobe walls. The performance of shear walls in tests of model buildings that had been retrofitted with vertical center-core rods was substantially better than that of walls that were not modified. The allowable shear design levels for walls could be increased by inclusion of vertical center-core rods at intervals of 1–2 m (3–7 ft.) on center. An associated design stress increase of 50%–100% would seem justified, based upon the results obtained on the models in the dynamic tests.

Cables, Straps, and Center-Core Rods

Cables and straps can be used to strengthen and add ductility to an unreinforced masonry structural system. Pre-tensioning the cables is not required, and straps should be tightened to eliminate slack in the system. The purpose of the straps and cables is to provide (1) limitations on the relative displacement of cracked wall sections, (2) resistance to out-of-plane flexure, and (3) in-plane continuity.

Center-core elements can be used to prevent the type of corner failure illustrated in the previous chapter (see fig. 7.6). Any location where a door or window is very close to a corner may be vulnerable to collapse during an earthquake. The selective use of grouted center-core elements that have been properly attached to the roof system or bond-beam elements should be effective in preventing extensive damage in all types of adobe buildings.

Case Study 1: Rancho Camulos

The full implementation of a complete, stability-based retrofit system can be seen in the main residence at the Del Valle Adobe at the Rancho Camulos Museum near Piru, in Ventura County, California, approximately 50 miles north of Los Angeles (Ginell and Tolles 1999).

The Del Valle Adobe was originally known as Rancho San Francisco, a ranch of Mission San Fernando. Through the years, the structure evolved into a U-shaped complex around a central courtyard or patio. The Historic American Buildings Survey dated the earliest portion of the building to 1841 and the additions to about 1846 (HABS: CA-38), but recent publications provide contradictory information regarding these dates (Delong 1980; Smith 1977).

The Rancho Camulos–Del Valle Adobe complex is considered to be a prime example of California's old ranchos because many of the elements typical of these structures—such as the *cocina* (kitchen), chapel, and winery—have survived essentially in their original form. It is undoubtedly the most famous of the surviving ranchos, having been identified as the home of the heroine of *Ramona*, a well-known romance novel by Helen Hunt Jackson set in early California. Its significance as a national symbol of the early days of California is difficult to overstate. Rancho Camulos has been designated as a National Historic Landmark.

A plan of the building complex is shown in figure 8.3. Figure 8.4 is a drawing of the building without the roof, showing the extent of the damage after the 1994 Northridge earthquake. The need for seismic retrofitting of the structure was specified in the historic structure report that was prepared prior to finalization of the design of the structural modifications. Preparation of the HSR was funded by the California Heritage Fund.

The south part of the building complex is a one-and-a-half-story, *tapanco*-style structure that is largely original. The northwest corner and west wings of the building are one story in height, as is the *cocina*, which is located in the north side of the complex. "Ramona's room," in the southeast corner of the building, is a single-story room that is attached to the one-and-a-half-story section. The building was damaged extensively during the Northridge earthquake. Two walls of Ramona's room collapsed (fig. 8.5, see fig. 5.18c) and the adjacent gable-end wall was severely damaged, but did not collapse. The bedroom in the southwest corner collapsed (fig. 8.6, see fig. 1.4a). Crack damage was widespread throughout the building, and adobe spallation and collapse was common, especially in water-weakened areas (Tolles et al. 2000).

The retrofit system used at the Del Valle Adobe consisted of stainless-steel cables and a partial plywood diaphragm for horizontal elements; flat, woven-nylon straps for vertical-wall reinforcement on existing walls; and center-core rods in the reconstructed walls, which were sometimes 30 cm and sometimes 60 cm thick (12 and 24 in.).

Cables were recommended for use as upper-wall elements because of their greater strength and stiffness than nylon straps. Horizontal elements may be subjected to large loads and over a longer

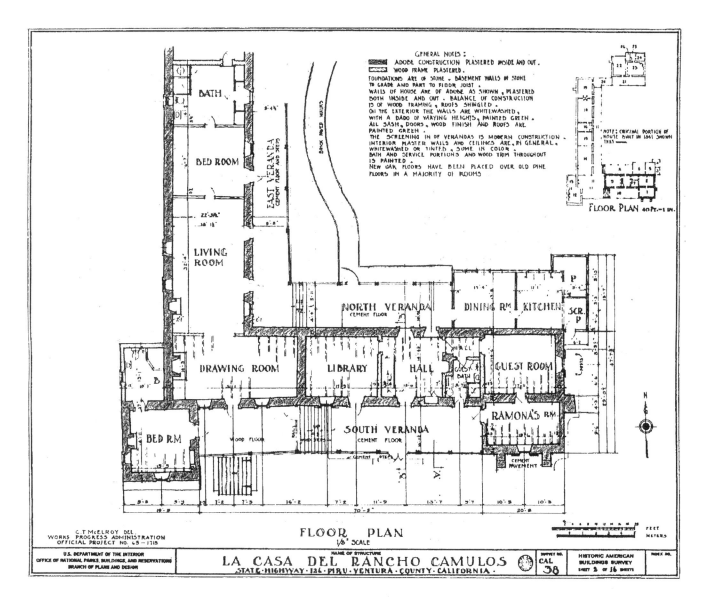

GENERAL NOTES:

☒ ADOBE CONSTRUCTION PLASTERED INSIDE AND OUT.
☐ WOOD FRAME PLASTERED.
FOUNDATIONS ARE OF STONE. BASEMENT WALLS OF STONE
TO GRADE AND PART TO FLOOR JOIST.
WALLS OF HOUSE ARE OF ADOBE AS SHOWN, PLASTERED
BOTH INSIDE AND OUT. BALANCE OF CONSTRUCTION
IS OF WOOD FRAMING, ROOFS SHINGLED.
ON THE EXTERIOR THE WALLS ARE WHITEWASHED,
WITH A DADO OF VARYING HEIGHTS, PAINTED GREEN.
ALL SASH, DOORS, WOOD FINISH AND ROOFS ARE
PAINTED GREEN.
THE SCREENING IN OF VERANDAS IS MODERN CONSTRUCTION.
INTERIOR PLASTER WALLS AND CEILINGS ARE, IN GENERAL,
WHITEWASHED OR TINTED, SOME IN COLOR.
BATH AND SERVICE PORTIONS AND WOOD TRIM THROUGHOUT
IS PAINTED.
NEW OAK FLOORS HAVE BEEN PLACED OVER OLD PINE
FLOORS IN A MAJORITY OF ROOMS.

NOTE: ORIGINAL PORTION OF
HOUSE BUILT IN 1841 SHOWN
THUS ——

FLOOR PLAN 40 FT.=1 IN.

FLOOR PLAN
1/8" SCALE

C.T. McELROY DEL.
WORKS PROGRESS ADMINISTRATION
OFFICIAL PROJECT NO. 65 - 1715

U.S. DEPARTMENT OF THE INTERIOR
OFFICE OF NATIONAL PARKS, BUILDINGS, AND RESERVATIONS
BRANCH OF PLANS AND DESIGN

LA CASA DEL RANCHO CAMULOS
STATE·HIGHWAY·126·PIRU·VENTURA·COUNTY·CALIFORNIA·

SURVEY NO.
CAL
38

HISTORIC AMERICAN
BUILDINGS SURVEY
SHEET 2 OF 16 SHEETS

INDEX NO.

Figure 8.3

Floor plan of Del Valle Adobe, Rancho Camulos Museum, Piru, Calif. (courtesy Historic American Buildings Survey).

length than vertical loads, and as a result, a high degree of stiffness would be advantageous. Vertical elements were required to conform to small-radius bends, and therefore greater flexibility was needed.

Vertical wall straps were installed on transverse shear walls to ensure ductility of these structural elements. Some of the existing interior walls were 30 cm (12 in.) thick and 2.7–3.3 m (9–11 ft.) high (S_L = 9–11); others were 60 cm (24 in.) thick (S_L = 6) but had suffered severe in-plane damage or had been modified earlier by door and window

Figure 8.4

Earthquake damage to Del Valle Adobe (Tolles et al. 1996).

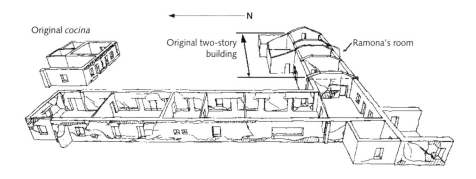

Figure 8.5
Collapsed wall of Ramona's room, Del Valle Adobe.

Figure 8.6
Collapsed southwest bedroom wall of Del Valle Adobe.

alterations. Vertical center-core rods, coupled with horizontal ladder trusses, were used on newly reconstructed walls that had collapsed during the 1994 Northridge earthquake.

Galvanized materials were not used in locations where contact with the fresh lime mortar or stucco might occur. The highly alkaline lime can interact chemically with wet galvanized materials, and therefore stainless steel was selected for steel elements located on or near the surfaces of the adobe walls.

The end connections for cables and straps were designed to distribute stresses and were important to ensure the ductility of these structural elements. Stresses in these elements can result in some localized damage to the adobe material, and this should be held to a minimum.

The end connections for the cables used at the Del Valle Adobe are shown in the schematic drawing in figure 8.7. The cables made straight runs and were terminated by threaded sections that were bolted to steel end plates. The plates were mounted over a layer of wire mesh; this helped reduce localized stresses in the adobe. Crossties made of heavy, solid nylon, electronic-type "cable ties" were installed through the wall along the lengths of the cables at intervals of about 1 m (3.3 ft.) on center to tie parallel cables to each other.

Details of the vertical straps used at the Del Valle Adobe are shown in a schematic drawing in the previous chapter (see fig. 7.3). The straps were routed through a hole drilled at the base of the wall and over the wood plate on top of the wall and then were tightened using two sets of D-rings mounted on one face of the wall. The hole at the base of the wall was lined with a section of plastic pipe measuring 3.8 cm (1.5 in.) in diameter, to reduce the extent of abrasion damage to the adobe during installation and when large dynamic earthquake loads are applied.

Straps and cables were placed in 2.5 cm (1 in.) chases cut into the adobe wall stucco. The

Figure 8.7
Horizontal cable-end connections used in the retrofit of Del Valle Adobe.

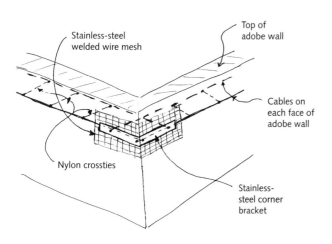

Stainless-steel welded wire mesh

Top of adobe wall

Cables on each face of adobe wall

Nylon crossties

Stainless-steel corner bracket

chases were then covered with stainless-steel wire mesh and lime stuccoed to match the existing wall surfaces.

The design of the retrofitting elements was based on a combination of engineering judgment and physical test data. Loads in cables and straps were measured during the large-scale GSAP tests, which were based on a prototype building with adobe walls 0.4 m (16 in.) thick. The peak stress in the horizontal element was 286 kg (631 lbs.) and was only 55 kg (120 lbs.) in the vertical elements. The allowable loads for both the cable and the straps were well above these levels. A commonly used stainless-steel cable, 13 mm (0.5 in.) in diameter, was selected for cost effectiveness. The flat, woven nylon strap, 3.8 cm (1.5 in.) in width, that was used had a strength of 1364 kg (3000 lbs.), and this was more than adequate.

The horizontal cable was attached to the second-floor framing in areas of the building with a second floor. The peak load in the horizontal cables during the large-scale GSAP tests was found to be 172 kg (380 lbs.).

Interior walls were retrofitted with horizontal rods that were connected to the horizontal cables on the exterior walls. The rods were combined with the vertical straps, and this produced a flexible support system for the thin interior walls, many of which had slenderness ratios between 9 and 10.

Nylon crossties that connected the exterior and interior vertical straps at mid-wall height remained intact and did not fail during the GSAP tests. Peak stresses in the crossties would be significantly less than those measured in the straps at the second-floor level. The solid nylon cable ties used in the tests were rated for working stress levels of 1.7 MPa (247 psi).

Repair of the main residence at Rancho Camulos included complete reconstruction of four sections of exterior walls, repair of severely cracked walls throughout the building, and replacement of about 15%–20% of the plaster that either fell off the adobe walls or was too loose to repair. The total cost of the repair and retrofitting with straps, cables, and center-core rods was about $463 per square meter ($43 per sq. ft.); the cost of the retrofit implementation alone was approximately $301 per square meter ($28 per sq. ft.).

Case Study 2: Casa de la Torre

The second case study is concerned with Casa de la Torre, an historic adobe building located in the National Landmark District of Monterey, California. Although the building had not sustained recent earthquake damage, it was seismically retrofitted at the request of the owner. The City of Monterey provided some financial support to encourage the participation of qualified professional specialists in the conservation work. As in the case of the Del Valle Adobe retrofitting, an historic structure report was prepared prior to design finalization. The HSR preparation, in this instance, was sponsored by the City of Monterey Historic Preservation Commission.

Francisco Pinto constructed Casa de la Torre in 1851–52. In 1862, Jose de la Torre purchased the then one-and-a-half-story adobe dwelling, which was occupied by the de la Torre family until 1923. In 1924, artist Myron Oliver bought the building and remodeled the adobe structure extensively. He eliminated the upper half story and created a studio by installing a tall arched window in the north wall. The interior of the studio was designed to resemble a mission chapel, with redwood cathedral ceilings, a choir loft, and "River of Life" carved doors. He also replaced the existing shingled roof with a Mission Revival–style tile roof.

As of this writing, in addition to seismic retrofitting, Casa de la Torre was undergoing rehabilitation to reflect its 1924 reintegration by Oliver, who was the father of the historic preservation movement in Monterey. A plan drawing of Casa de la Torre is shown in figure 8.8 and photographs of the lower roof (west side) and the window wall and porch (north and east sides) are shown in figures 8.9 and 8.10, respectively. The dimensions of the east and west walls of the main section are 11.3 × 4.3 × 0.6 m (37 × 14 × 2 ft.). The end walls, which are 6.1 m (20 ft.) wide, are gabled, and a loft covers about 30% of the south end of the main section.

The area most vulnerable to seismic damage appeared to be the north gable-end wall containing a large glass window. To strengthen and stabilize this wall, two center-core rods, 1.9 cm (0.75 in.) in diameter, were inserted vertically in the wall on each side of the window. The upper ends of the rods were threaded and bolted to the upper wood wall plate. The rods, which extended only to the top of the stone foundation, were grouted in place using a high-strength, cementitious grout. Center-core rods were also placed in each pier of the east wall (see fig. 8.8) because the freestanding porch roof provided little or limited support for this wall.

Because the walls on the west and south sides were braced by the single-story roof at about 1.2–1.6 m (4–6 ft.) from the tops of these walls, center-core rod installation was not considered necessary, except for the placement of a single core rod in the first pier adjacent to the

Figure 8.8
Floor plan of Casa de la Torre, Monterey, Calif.

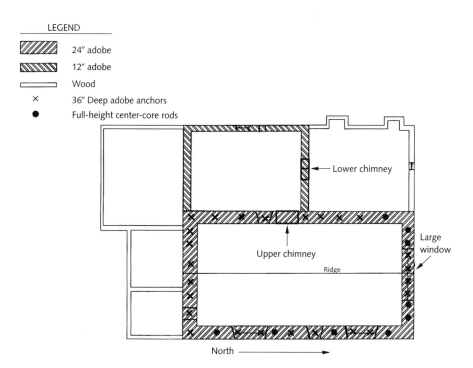

LEGEND

▨▨▨	24″ adobe
▨▨▨	12″ adobe
▭	Wood
×	36″ Deep adobe anchors
●	Full-height center-core rods

Lower chimney

Upper chimney

Ridge

Large window

North ⟶

Figure 8.9
Casa de la Torre, west side view.

Figure 8.10
Casa de la Torre, north (window) and east walls (photo courtesy
Tony Crosby).

north wall. This rod was added to strengthen the corner and to resist the
forces that might be generated if the north wall tended to move outward.
Short roof anchor rods, 1.6 cm in diameter by 91 cm long (0.625×36
in.), were embedded at intervals along the tops of the adobe walls, and
these too were grouted in place using the cementitious grout.

The cost of this limited project to retrofit the high-walled
section of the Casa de la Torre was about $25,000 or roughly $364 per
square meter ($34 per sq. ft.).

Summary of Retrofit Considerations for Adobe Buildings with Walls of Different Slenderness Ratios

As has been discussed, a highly significant element affecting the retrofit
design for a historic adobe structure is the slenderness ratio. The following outlines some of the stability-based design considerations for retrofitting buildings with thick, moderate, or thin walls.

Design example: Thick-walled buildings ($S_L < 6$)

Lateral load distribution
Attachment of the tops of walls to the roof system is required to prevent
overturning by rotation about the base. Shear walls will be subjected to
much lower load levels than the values calculated using a typical static
design procedure because of the out-of-plane resistance to overturning of
thick walls. A reduction of calculated shear forces distributed to the in-
plane walls can probably be justified.

Vertical reinforcement for walls
In most instances, vertical wall reinforcements may not be required
because the wall displacement would need to be very large before stability problems would occur. Nevertheless, if the goal is to limit permanent
offsets during severe ground motions, center-core elements could be used.

Design example: Moderately thick-walled buildings ($S_L = 6-8$)

Lateral load distribution
Attachment of the tops of walls to the roof system is required to prevent wall overturning as a result of rotation about the base. Some minor strengthening of the roof or floor systems will add a degree of redundancy to the structural system. Shear walls will be subjected to lower load levels than calculated using typical static design methods due to the out-of-plane overturning resistance of these walls. Here again, reduced calculated shear forces to the in-plane walls may be justified.

Vertical reinforcement for walls
Vertical reinforcement may be required on these walls to increase the ductility of the walls or to reduce the size of permanent offsets. Vertical straps will increase ductility, while center-core rods will both increase ductility and reduce permanent offsets.

Design example: Thin-walled buildings ($S_L > 8$)

Lateral load distribution
Thin walls have little-to-no resistance to overturning and will freely rotate about their bases. A roof or attic diaphragm system will be required that can transfer forces to in-plane shear walls.

Vertical reinforcement for walls
Vertical reinforcements *must* be used on thin walls to ensure that they will perform adequately in the out-of-plane direction. Vertical straps can provide sufficient safety in most situations, but a more secure solution involves addition of center-core rods, which will increase both ductility and strength while preventing degradation during extended seismic shaking.

Moisture-damaged adobe
All adobe walls should be inspected to detect water damage, especially near the base of the walls. The stability advantage of thick walls is compromised when adobe bricks at the base of the wall have been damaged by prior wet-dry cycling and when the wall contains excessive moisture. It is particularly important to examine the raw, unplastered faces of walls that have been rendered with Portland cement–based materials, which do not allow the rapid evaporation of water. Loss of load-carrying capability and subsequent wall collapse are highly likely in such cases of moisture damage. Structural repairs are mandatory, but they should not be carried out until the source of the water has been eliminated and the wall has dried. Deteriorated adobe bricks should be removed and replaced by new adobe bricks and mortar. If the source of water damage cannot be eliminated, the new bricks must be fabricated from a stabilized type of adobe that resists deterioration on contact with water.

Chapter 9

Conclusions

Analysis of the results of seismic events in recent years, particularly the Northridge earthquake of 1994, suggests that failure to retrofit historic adobe buildings will continue to result in serious losses. Acceptance of the retrofit challenge will produce long-term benefits in terms of both preserving historic resources and assuring life safety. Neither loss of historic fabric owing to overly invasive retrofit strategies nor direct fabric destruction by an earthquake is desirable. A balance can be achieved whereby the authenticity of a historic building and public safety are ensured, and these guidelines are designed to provide information than can help achieve a seismic retrofit strategy consistent with conservation principles.

In addition to cultural losses, earthquakes can cause adverse economic effects on historic adobe tourist destinations, such as the California missions. Recent state tourism research indicates that historic sites rank immediately after natural wonders in visitor popularity. The California missions are among the most visited of such tourist destinations in California (Murphy 1992).

The architecture of the historic adobes and early Spanish missions of California is associated with the state's identity in the world—a result of intense activity that started in the nineteenth century to promote the state as an idyllic region in which to live or to visit. Responsibility lies with this and succeeding generations to preserve what remains of these structures by safeguarding them and their occupants from earthquake damage.

In attempting to meet the challenge of preserving and protecting historic adobe buildings as examples of New World architectural antiquities, it is important to follow judicious planning procedures, regardless of budget size. Consideration of all the relevant issues in the planning phase will yield rewards proportionate to the effort expended. There is a chance that, due to the lack of requisite information about historical and architectural significance, important features for which an owner or manager is responsible and held accountable (if only by history) may be inadvertently lost or altered to the extent that authenticity is substantially diminished.

Of course, there can be no guarantee that following these guidelines will ensure satisfaction with the performance of a design team or that a specific project will be favorably reviewed by a historic resources commission. Employment of an architect as project supervisor does not ensure project approval, but it generally improves the chances of overall project success. Nor can it be claimed responsibly that seismic retrofitting will prevent a building from being damaged by seismic activity. However, some measure of both public safety and damage control can be obtained while preserving a significant portion of the authenticity of the building.

The minimally invasive retrofitting designs outlined here have been evaluated experimentally and found to confer stability on the model adobe structures tested on a seismic simulator. It should not be inferred that the designs described here are unique or that alternate designs may not work as well. Nonetheless, we feel that the principle of seeking to provide seismic stability—rather than improving the strength of an existing adobe building—has been demonstrated and should be considered when designing future retrofits for historic adobe structures.

As Jokilehto (1985) noted, the concern for preserving cultural heritage has been expressed, with a few exceptions, since antiquity. Concepts and procedures change with time, however, and it is prudent to develop a well-thought-out conservation policy that uses a case-by-case approach, rather than blindly following so-called established precepts.

Getty Seismic Adobe Project

The objective of the Getty Seismic Adobe Project (GSAP) was to contribute to the body of knowledge about the earthquake behavior of historic adobe buildings by developing an understanding of failure modes and by developing technical procedures for improving the seismic performance of existing monumental adobe structures consistent with maintaining architectural, historic, and cultural values.

The primary accomplishments of this conservation project were the formulation of a general theoretical framework for understanding the dynamic performance of historic adobe buildings during seismic activity, the development of a methodology for designing retrofit systems for historic adobe buildings, and the presentation of data on a set of experimentally verified retrofit measures that could be used for the stabilization of these structures.

The final result of this effort is not a step-by-step design manual but one that requires study and the application of engineering judgment. Designing a retrofit system for an unreinforced adobe building is part science and part art and requires an understanding of the strengths and weaknesses of the adobe material.

Overall Structure of Project

The work carried out during GSAP was divided into three phases:

Phase 1: Evaluation of existing knowledge and practices concerning seismic stabilization of historic adobe buildings and the development of a technical foundation on which methods for improving their seismic resistance could be based.

Phase 2: Initiation of research necessary to validate the retrofitting concepts and to supplement what is currently known. Research included shaking-table tests as well as analytical modeling. The occurrence of the 1994 Northridge earthquake in the Los Angeles area proved to be a unique opportunity to study the effects of strong earthquake motions on existing historic adobe buildings. This research was added to the original plan for Phase 2.

Phase 3: A documentation phase, which included preparation and distribution of research reports, journal articles, presentations at technical conferences, and the publication of this book, consisting of guidelines for the planning and retrofitting of historic adobe structures. The planning sections discuss relevant issues concerning historic adobe structures, conservation, and cultural values. They also provide an outline of the steps to take when planning the seismic retrofit of a historic adobe building. The engineering aspect of the book offers a theoretical basis for understanding the seismic performance and technical procedures used for designing seismic retrofit measures for historic adobes.

It was felt that, to be useful, these guidelines would need the wide professional support of the technical community. To achieve that, they would have to be workable in application and responsive to real seismic retrofit problems. The decision was made, therefore, to approach GSAP as a cooperative endeavor by a group of individuals who were experts in the analysis of adobe's seismic behavior and who were familiar with the many complex cultural issues that influence the possible modification of historic adobe buildings.

GSAP benefited from the advice of an advisory committee that was appointed to assure that the project was proceeding in a logical way to achieve its objectives. The GSAP Advisory Committee had two principal responsibilities:

- To monitor project activities and advise the project manager on the management and direction of GSAP, and
- To review the technical activities and accomplishments of GSAP and advise the project director and the project manager on its findings.

Advisory Committee and Project Personnel

GSAP Advisory Committee members
Edward E. Crocker, architectural conservator and contractor, Santa Fe, New Mexico
Anthony Crosby, historical architect, formerly with the National Park Service, Denver, Colorado
M. Wayne Donaldson, historical architect, San Diego, California
Melvyn Green, seismic structural engineer, Torrance, California
James Jackson, architect, California State Parks, Sacramento, California
Helmut Krawinkler, professor, Structural Engineering, Stanford University, Palo Alto, California
John Loomis, architect, Thirtieth Street Architects, Newport Beach, California

Nicholas Magalousis, professor, Santa Ana College, and
    former curator, Mission San Juan Capistrano, San Juan
    Capistrano, California
Julio Vargas Neumann, professor, Structural Engineering,
    Pontifica Universidad Catolica del Peru, Lima, Peru

GSAP personnel
Neville Agnew, former GSAP director, Getty Conservation
    Institute
William S. Ginell, GSAP director and materials scientist,
    Getty Conservation Institute
Edna E. Kimbro, architectural historian and conservator
Charles C. Thiel Jr., seismic engineer
E. Leroy Tolles, principal investigator and seismic engineer
Frederick A. Webster, seismic engineer

Summary of GSAP Activities

The activities of GSAP included research, testing, and field investigations.
Consultation with the members of the Advisory Committee and other
professionals increased the relevance of the GSAP efforts to actual seismic damage problems at historic sites. The final results of these efforts
were interim technical reports, a final report on the research studies
(Tolles et al. 2000), presentations at technical conferences, technical journal articles, a survey of earthquake damage to historic adobe buildings
after the Northridge earthquake (Tolles et al. 1996), and these guidelines.
        Phase 1 included a review of existing retrofitting practices, a
literature review, and the preliminary development of the planning
guidelines (Thiel et al. 1991). After the Phase 1 research and discussions
with the Advisory Committee, a research program was outlined. The
research effort included shaking-table tests at Stanford University on
one-fifth-scale model adobe structures. Tests on the first three model
buildings were detailed in the report on second-year activities of GSAP
(Tolles et al. 1993).
        Following the initial tests, the Northridge earthquake occurred
in the Los Angeles area. Although it was unfortunate that so many buildings were damaged, this event was extremely beneficial for the research
effort. A great deal of previously undocumented, detailed information on
historic building earthquake damage was collected. The seismic shaking-
table research effort on six additional small-scale models was completed at
Stanford University after the Northridge earthquake. Following the small-
scale tests, studies of two one-half scale models were carried out on a large
shaking table in Skopje, the Republic of Macedonia.
        The following is a chronology of the principal GSAP activities.

  • 1991–92 Phase 1, research and preliminary Advisory
        Committee meeting

- 1991   Report of first-year activities (Thiel et al. 1991)
- 1992   Tests on models 1, 2, and 3, simple 1:5-scale adobe models
- 1993   Report of second-year activities (Tolles et al. 1993)
- 1994   Tests on models 4, 5, and 6, simple 1:5-scale adobe models
- 1994–95 Survey and report on the damage to historic adobe buildings resulting from the 1994 Northridge earthquake (Tolles et al. 1996)
- 1994   Test on model 7, *tapanco*-style, 1:5-scale, retrofitted adobe model
- 1995   Tests on models 8 (retrofitted) and 9 (unretrofitted control), *tapanco*-style, 1:5-scale models with moderately thick walls
- 1996   Tests on 1:2-scale models, models 10 and 11 (Gavrilovic et al. 1996)
- 2000   Final report summarizing all test activities (Tolles et al. 2000)
- 2001   Report of third-year activities: shaking-table tests of large-scale adobe structures (Ginell et al. 2001)
- 2002   This volume, summarizing and synthesizing the important aspects of the research effort

Appendix B

The Unreinforced Masonry Building Law, SB547

Following is chapter 12.2 of the Unreinforced Masonry Building Law, SB547, of the Seismic Safety Commission (2000).

Chapter 12.2 Building Earthquake Safety
Chapter 12.2 was added by Stats. 1986, c. 250, § 2.

§ 8875. Definitions. Unless the context otherwise requires, the following definitions shall govern the construction of this chapter:

(a)   "Potentially hazardous building" means any building constructed prior to the adoption of local building codes requiring earthquake resistant design of buildings and constructed of unreinforced masonry wall construction. "Potentially hazardous building" includes all buildings of this type, including, but not limited to, public and private schools, theaters, places of public assembly, apartment buildings, hotels, motels, fire stations, police stations, and buildings housing emergency services, equipment, or supplies, such as government buildings, disaster relief centers, communications facilities, hospitals, blood banks, pharmaceutical supply warehouses, plants, and retail outlets. "Potentially hazardous building" does not include any building having five living units or less. "Potentially hazardous building" does not include, for purposes of subdivision (a) of Section 8877, any building which qualifies as "historical property" as determined by an appropriate governmental agency under Section 37602 of the Health and Safety Code.

(b)   "Local building department" means a department or agency of a city or county charged with the responsibility for the enforcement of local building codes.

§ 8875.1 Establishment of program; identification of potentially hazardous buildings; advisory report

A program is hereby established within all cities, both general law and chartered, and all counties and portions thereof located within seismic zone 4, as defined and illustrated in Chapter 2-23 of Part 2 of Title 24 of the California Administrative Code, to identify all potentially hazardous buildings and to establish a program for mitigation of identified potentially hazardous buildings.

By September 1, 1987, the Seismic Safety Commission, in cooperation with the League of California cities, the County Supervisors Association of California, and California building officials, shall prepare an advisory report for local jurisdictions containing criteria and procedures for purposes of Section 8875.2.

(Formerly § 8876, added by Stats. 1986, c. 250, § 2. Renumbered § 8875.1 and amended by Stats. 1987, c. 56, § 62)

§ 8875.2 Local building departments; participation in mitigation programs; reports

Local building departments shall do all of the following:

(a)   Identify all potentially hazardous buildings within their respective jurisdiction on or before January 1, 1990. This identification shall include current building use and daily occupancy load. In regard to identifying and inventorying the buildings, the local building departments may establish a schedule of fees to recover the costs of identifying potentially hazardous buildings and carrying out this chapter.

(b)   Establish a mitigation program for potentially hazardous buildings to include notification to the legal owner that the building is considered to be one of a general type of structure that historically has exhibited little resistance to earthquake motion. The mitigation program may include the adoption by ordinance of a hazardous buildings program, measures to strengthen buildings, measures to change the use to acceptable occupancy levels or to demolish the building, tax incentives available for seismic rehabilitation, low-cost seismic rehabilitation loans available under Division 32 (commencing with Section 5506) of the Health and Safety Code, application of structural standards necessary to provide for life safety above current code requirements, and other incentives to repair the buildings which are available from federal, state, and local programs. Compliance with an adopted hazardous buildings ordinance or mitigation program shall be the responsibility of building owners.

Nothing in this chapter makes any state building subject to a local building mitigation program or makes the state or any local government responsible for paying the cost of strengthening a privately owned structure, reducing the occupancy, demolishing a structure, preparing engineering or architectural analysis, investigation, or design, or other costs associated with compliance of locally adopted mitigation programs.

(c)   By January 1, 1990, all information regarding potentially hazardous buildings and all hazardous building mitigation programs shall be reported to the appropriate legislative body of a city or county and filed with the Seismic Safety Commission.

§ 8875.3 Local jurisdictions; immunity from liability

Local jurisdictions undertaking inventories and providing structural evaluations of potentially hazardous buildings pursuant to this chapter shall have the same immunity from liability for action or inaction taken pursuant of this chapter as is provided by Section 19167 of the Health and Safety Code for action or failure to take any action pursuant to Article 4 (commencing with Section 19160) of Chapter 2 of Part 3 of Division 13 of the Health and Safety Code.

§ 8875.4 Annual report

The Seismic Safety Commission shall report annually, commencing on or before June 30, 1987, to the Legislature on the filing of mitigation programs from local jurisdiction. The annual report required by this section shall review and assess the effectiveness of building reconstruction standards adopted by cities and counties pursuant to this article and shall supersede the reporting requirement pursuant to Section 19169 of the Health and Safety Code.

§ 8875.5 Coordination of responsibilities

The Seismic Safety Commission shall coordinate the earthquake-related responsibilities of government agencies imposed by this chapter to ensure compliance with the purposes of this chapter.

§ 8875.6 Transfer of unreinforced masonry building with wood frame floors or roofs; duty to deliver to purchaser earthquake safety guide

On and after January 1, 1993, the transferor, or his or her agent, of any unreinforced masonry building with wood frame floors or roofs, built before January 1, 1975, which is located within any county or city will, as soon as practicable before the sale, transfer, or exchange, deliver to the purchaser a copy of the *Commercial Property Owner's Guide to Earthquake Safety* described in Section 10147 of the Business and Professions Code. *This section shall not apply to any transfer described in Section 8893.3.*

§ 8875.7

If the transferee has received notice pursuant to Section 8875.8, and has not brought the building or structure into compliance within five years of that date, the owner shall not receive payment from any state assistance program for earthquake repairs resulting from damage during an earthquake until all other applicants have been paid.

§ 8875.8

(a)   Within three months of the effective date of the act amending this section, enacted at the 1991–92 Regular Session, any owner who has received actual or constructive notice that a building located in seismic zone 4 is constructed of unreinforced masonry shall post in a conspicuous place at the entrance of the building, on a sign not less than 5 × 7 inches, the following statement, printed in not less than 30-point bold type:

> This is an unreinforced masonry building. Unreinforced masonry buildings may be unsafe in the event of a major earthquake.

(b)   Notice of the obligation to post a sign, as required by subdivision (a), shall be included in the *Commercial Property Owner's Guide to Earthquake Safety*.

§ 8875.9
Section 8875.8 shall not apply to unreinforced masonry construction if the walls are non-load-bearing with steel or concrete frame.

§ 8875.95
No transfer of title shall be invalidated on the basis of failure to comply with this chapter.

Reprinted courtesy of the Seismic Safety Commission, Sacramento, Calif.

California Building Code and Seismic Safety Resources

State Historical Building Code

The California State Historical Building Code (SHBC), revised in 1999, is available from the International Conference of Building Officials (ICBO), 5360 Workman Mill Road, Whittier, CA 90601-2298. To order, call (800) 284-4406 or visit the website at www.icbo.org.

The provisions of the code applicable to adobe masonry are contained in chapter 8-8, section 8-806. Chapter 8-1, "Administration," section 8-104, deals with reviews and appeals.

The executive director of the State Historical Building Safety Board may be contacted c/o Division of the State Architect, 1130 K Street, Suite 101, Sacramento, CA 95814; (916) 445-7627.

Seismic Safety Commission

The Seismic Safety Commission is located at 1755 Creekside Oaks Drive, Suite 100, Sacramento, CA 95833; (916) 263-5506; www.seismic.ca.gov.

Publications related to earthquake safety and retrofits available from the Seismic Safety Commission include the following (many available online in PDF format):

> "Architectural Practice and Earthquake Hazards: A Report of the Committee on the Architect's Role in Earthquake Hazard Mitigation." Seismic Safety Commission Report no. SSC 91-10.
>
> "Guidebook to Identify and Mitigate Seismic Hazards in Buildings." Seismic Safety Commission Report no. SSC 87-03. December 1987.
>
> "Status of the Unreinforced Masonry Building Law." 2000 Biennial Report to the Legislature. Seismic Safety Commission Report no. SSC 00-02.
>
> Earthquake risk management tools for decision makers:
> - "A Guide for Decision Makers." Publication no. SSC 99-06.
> - "Mitigation Success Stories." Publication no. SSC 99-05.
> - "A Toolkit for Decision Makers." Publication no. SSC 99-04.

Following are some sources of historic structure report formats and methodology:

> *Cultural Resource Management* vol. 13: nos. 4 and 6, 1990. These two issues of this National Park Service technical bulletin are devoted to historic structure reports. They are available through the government documents section of large libraries (contact your local reference librarian) or in PDF format online at crm.cr.nps.gov.

> "Heritage Recording," *APT Bulletin: The Journal of Preservation Technology* vol. 22, nos. 1 and 2, 1990. Available via interlibrary loan at local public libraries.

> "Historic Structure Report." New guidelines for preparation of historic structure reports issued in 2002. Available from American Society for Testing and Materials (ASTM), 100 Bar Harbor Drive, P.O. Box C700, West Conshohocken, PA 19428-2959; www.astm.org.

> Historic Structure Report Format. This simple outline is revised periodically and is available from the Office of Historic Preservation, California State Parks, P.O. Box 942896, Sacramento, CA 94296-0001.

> "Historic Structure Reports: Special Issue." *APT Bulletin: The Journal of Preservation Technology*, volume 28, no. 1, 1997. Eleven papers on the use of historic structure reports.

> "Preparing a Historic Structure Report," NPS-28 Cultural Resource Management Guideline, July 1994, Director's Order no. 28. National Park Service, U.S. Department of the Interior, Washington, D.C. Available through government documents section of large libraries (contact your local reference librarian).

Sources of Information and Assistance

The following organizations offer useful information and professional guidance of various kinds:

American Association for State and Local History (AASLH) provides "leadership and support for its members who preserve and interpret state and local history in order to make the past more meaningful to all Americans." Preservation information is included in the quarterly *History News*. AASLH also publishes a series of technical leaflets and special reports on pertinent topics, as well as other publications. AASLH, 1717 Church Street, Nashville, TN 37203-2991; (615) 320-3203; www.aaslh.org.

American Institute for Conservation of Historic and Artistic Works (AIC) publishes the brochure "Guidelines for Selecting a Conservator" and provides assistance in locating and selecting conservation professionals through the AIC Guide to Conservation Services. AIC also publishes the *Journal of the American Institute for Conservation* and the *AIC Directory*, a catalogue of members listed by specialty, name, and location. AIC, 1717 K Street, N.W., Suite 200, Washington, D.C. 20006; (202) 452-9545; aic.stanford.edu.

American Institute of Architects (AIA) publishes a "Guide to Historic Preservation" that is available online in PDF format at www.aia.org/pia/hrc/8752PreservationGuide.pdf. The AIA also has a Historic Resources Committee. AIA, 1735 New York Avenue, N.W., Washington, D.C. 20006; (800) 626-7300; www.aia.org. AIA San Francisco Chapter, 130 Sutter Street, Suite 600, San Francisco, CA 94104; (415) 362-7397; www.aiasf.org.

Association for Preservation Technology International (APT) publishes *APT Communique* (a quarterly newsletter), a directory of members, and the *APT Bulletin*, an important resource for technical preservation information. The voluminous proceedings of the two Seismic Retrofit of Historic Buildings conferences (Conference Workbook) are available from the Western Chapter. APT, 4513 Lincoln Ave., Suite 213, Lisle, IL 60532-1290; (630) 968-6400; www.apti.org. APT Western Chapter, 85 Mitchell Blvd., Suite 1, San Rafael, CA 94903; (415) 491-4088.

California Council for the Promotion of History (CCPH) was founded to foster the preservation, documentation, interpretation, and management of California's historical resources. CCPH publishes an informative newsletter, *California History Action*, and organizes an annual conference, among other activities. CCPH also publishes the Register of Professional Historians (available online in PDF format) as well as a directory of organizations in the state focusing on history. CCPH, California State University, Sacramento, 6000 J Street, Sacramento, CA 95819-6059; (916) 278-4296; www.csus.edu/org/ccph/index.htm.

California Mission Studies Association (CMSA) is dedicated to the study and preservation of California's Native American, Hispanic, and early American past. It publishes a newsletter that includes articles on preservation related to Hispanic-era missions, presidios, adobe buildings, and historical archaeological sites of the period, as well as a directory of members giving professional information. The annual CMSA conference often features a preservation workshop or presentations on preservation issues. CMSA, P.O. Box 3357, Bakersfield, CA 93385; www.ca-missions.org.

California Office of Historic Preservation (OHP), part of the Department of Parks and Recreation, is the lead historic preservation agency for California. The OHP staffs the State Historical Resources Commission and administers the National Register, California Register, and California State Landmarks programs, among others. The office provides preservation assistance, grant funding applications, tax credit certification, and historical designation status information. A variety of publications, including a regular newsletter relating to historic preservation and the April 2001 publication "Historic Preservation Incentives in California," are available from the OHP. OHP, P.O. Box 942896, Sacramento, CA 94296-0001; (916) 653-6624; ohp.parks.ca.gov.

The OHP's California Historical Resources Information System (CHRIS) maintains a referral list for historical resources consultants "who have satisfactorily documented that they meet the Secretary of Interior's Standards for that profession." The list includes historical archaeologists, historians, historical architects, and architectural historians. The CHRIS coordinator can be contacted at the above address or by telephone at (916) 653-9125.

California Preservation Foundation (CPF) publishes *California Preservation*, a quarterly newsletter, and reports on preservation issues including seismic retrofitting. CPF organizes preservation workshops and sponsors the annual California Preservation Conference in conjunction with the California Office of Historic Preservation. CPF, 1611 Telegraph Ave., Suite 820, Oakland, CA 94612; (510) 763-0972; www.californiapreservation.org.

California State Historical Building Safety Board (SHBSB) publishes the State Historical Building Code (SHBC), which is available from the International Conference of Building Officials (ICBO), 5360 Workman Mill Road, Whittier, CA 90601-2298; (800) 284-4406; www.icbo.org. The provisions of the code applicable to adobe masonry are contained in chapter 8-8, "Archaic Materials and Methods of Construction," section 8-806, "Adobe"; chapter 8-7, "Alternative Structural Regulations"; and chapter 8-1, section 8-104, "Appeals,

Alternative Proposed Design, Materials and Methods of Construction."
State Historical Building Safety Board, c/o Division of the State
Architect, 1130 K Street, Suite 101, Sacramento, CA 95814;
(916) 445-7627; www.dsa.dgs.ca.gov/SHBSB/shbsb_main.asp.

CRATerre-EAG, the International Centre for Earth
Construction, is in the School of Architecture at the University of
Grenoble. It offers a vast amount of information on earthen materials
and their use, in addition to training courses in the technology of earthen
building construction. CRATerre-EAG, F-38092 Villefontaine Cedex,
France; www.craterre.archi.fr.

Historic American Buildings Survey documents historic buildings in historical reports, photographs, and measured drawings. Many of
these are available online; for information go to www.cr.nps.gov/
habshaer/coll/index.htm. Hard copies of photographs and drawings can
be obtained from the Library of Congress, Prints and Photographs
Division, 101 Independence Ave., S.E., Washington, D.C. 20540-4730,
attn: Reference Section.

**ICCROM (International Centre for the Study of the
Preservation and Restoration of Cultural Property)**, through its GAIA
Program, has developed training courses and encouraged research and
information dissemination on the preservation of historic and culturally
significant earthen material buildings, including adobe. The organization
maintains an extensive library on architectural conservation. ICCROM,
Via di San Michele, 13. I-00153 Rome, Italy; +39 06 585531;
www.iccrom.org.

National Park Service (NPS) offers preservation assistance as
well as a series of informational publications, including Preservation
Tech Notes and Preservation Briefs. NPS is the lead historic preservation
agency nationally, administering federal programs, including the National
Register of Historic Places and the National Historic Landmarks program, as well as historic sites and monuments. Heritage Preservation
Services, NPS, 1849 C Street, N.W., NC330, Washington, D.C. 20240;
www2.cr.nps.gov/tps/index.htm.

National Trust for Historic Preservation publishes
Preservation magazine, *Forum News*, and *Preservation Journal*, as well
as an Information Series and publications on specific preservation topics.
The National Trust also administers several grant and loan programs and
provides preservation assistance and information, among other activities.
National Trust for Historic Preservation, 1785 Massachusetts Ave.,
N.W., Washington, D.C. 20036; (202) 588-6000; www.nthp.org.

Partners for Sacred Places publishes a fund-raising guidebook
for religious properties, produces a regular newsletter, and holds a conference series (Sacred Trusts) on the stewardship of America's older and
historic religious buildings and sacred sites. The organization also has an
online database of resources and a program for professionals engaged in
restoring historic religious properties. Partners for Sacred Places, 1700
Sansom Street, Tenth Floor, Philadelphia, PA 19103; (215) 567-3234;
www.sacredplaces.org.

Register of Professional Archaeologists (RPA) is a directory of archaeologists "who have agreed to abide by an explicit code of conduct and standards of research performance, who hold a graduate degree in archaeology, anthropology, art history, classics, history, or another germane discipline and who have substantial practical experience." The searchable online directory is updated quarterly; a print version is published once a year. RPA, 5024-R Campbell Blvd., Baltimore, MD 21236; (410) 933-3486; www.rpanet.org.

Society for Historical Archaeology (SHA) "promotes scholarly research and the dissemination of knowledge concerning historical archaeology" and "is specifically concerned with the identification, excavation, interpretation, and conservation of sites and materials on land and underwater." SHA publishes a quarterly journal, *Historical Archaeology*; a quarterly newsletter; and occasional special publications. SHA's Conference on Historical and Underwater Archaeology is convened every January. SHA, P.O. Box 30446, Tucson, AZ 85751-0446; (520) 886-8006; www.sha.org.

Southwestern Mission Research Center (SMRC) produces the *SMRC Newsletter*, which often contains preservation information. It is available from the Arizona State Museum, University of Arizona, P.O. Box 210026, Tucson, AZ 85721. The annual Gran Quivira conference is organized by the readership of historians, archaeologists, architects, conservators, and interested parties. SMRC, P.O. Box 213, Tumacacori, AZ 85640; (520) 558-2396.

Federal Standards for Treatment of Historic Properties

The U.S. Department of the Interior, through the National Park Service, has established standards that apply to the alteration of historic properties and has outlined criteria for determining the eligibility of such properties for inclusion in the National Register of Historic Places. Planners should be aware of these standards and criteria (quoted verbatim here) when considering designs for seismic damage mitigation alterations to historic properties.

The Secretary of the Interior's Standards

Title 36—Parks, Forests, and Public Property

CHAPTER I—NATIONAL PARK SERVICE, DEPARTMENT OF THE INTERIOR

PART 68—THE SECRETARY OF THE INTERIOR'S STANDARDS FOR THE TREATMENT OF HISTORIC PROPERTIES

§ 68.1 Intent.

The intent of this part is to set forth standards for the treatment of historic properties containing standards for preservation, rehabilitation, restoration and reconstruction. These standards apply to all proposed grant-in-aid development projects assisted through the National Historic Preservation Fund. 36 CFR part 67 focuses on "certified historic structures" as defined by the IRS Code of 1986. Those regulations are used in the Preservation Tax Incentives Program. 36 CFR part 67 should continue to be used when property owners are seeking certification for Federal tax benefits.

§ 68.2 Definitions.

The standards for the treatment of historic properties will be used by the National Park Service and State historic preservation officers and their staff members in planning, undertaking and supervising grant-assisted projects for preservation, rehabilitation, restoration and reconstruction. For the purposes of this part:

(a)   *Preservation* means the act or process of applying measures necessary to sustain the existing form, integrity and materials of an historic property. Work, including preliminary measures to protect and stabilize the property, generally focuses upon the ongoing maintenance and repair of historic materials and features rather than extensive replacement and new construction. New exterior additions are not within the scope of this treatment; however, the limited and sensitive upgrading of mechanical, electrical and plumbing systems and other code-required work to make properties functional is appropriate within a preservation project.

(b)   *Rehabilitation* means the act or process of making possible an efficient compatible use for a property through repair, alterations and additions while preserving those portions or features that convey its historical, cultural or architectural values.

(c)   *Restoration* means the act or process of accurately depicting the form, features and character of a property as it appeared at a particular period of time by means of the removal of features from other periods in its history and reconstruction of missing features from the restoration period. The limited and sensitive upgrading of mechanical, electrical and plumbing systems and other code-required work to make properties functional is appropriate within a restoration project.

(d)   *Reconstruction* means the act or process of depicting, by means of new construction, the form, features and detailing of a nonsurviving site, landscape, building, structure or object for the purpose of replicating its appearance at a specific period of time and in its historic location.

§ 68.3 Standards.

One set of standards—preservation, rehabilitation, restoration or reconstruction—will apply to a property undergoing treatment, depending upon the property's significance, existing physical condition, the extent of documentation available and interpretive goals, when applicable. The standards will be applied taking into consideration the economic and technical feasibility of each project.

(a)   *Preservation.*
   (1) A property will be used as it was historically, or be given a new use that maximizes the retention of distinctive materials, features, spaces and spatial relationships. Where a treatment and use have not been identified, a property will be protected and, if necessary, stabilized until additional work may be undertaken.
   (2) The historic character of a property will be retained and preserved. The replacement of intact or repairable historic materials or alteration of features, spaces and spatial relationships that characterize a property will be avoided.

(3) Each property will be recognized as a physical record of its time, place and use. Work needed to stabilize, consolidate and conserve existing historic materials and features will be physically and visually compatible, identifiable upon close inspection and properly documented for future research.

(4) Changes to a property that have acquired historic significance in their own right will be retained and preserved.

(5) Distinctive materials, features, finishes and construction techniques or examples of craftsmanship that characterize a property will be preserved.

(6) The existing condition of historic features will be evaluated to determine the appropriate level of intervention needed. Where the severity of deterioration requires repair or limited replacement of a distinctive feature, the new material will match the old in composition, design, color and texture.

(7) Chemical or physical treatments, if appropriate, will be undertaken using the gentlest means possible. Treatments that cause damage to historic materials will not be used.

(8) Archeological resources will be protected and preserved in place. If such resources must be disturbed, mitigation measures will be undertaken.

(b)   *Rehabilitation.*

(1) A property will be used as it was historically or be given a new use that requires minimal change to its distinctive materials, features, spaces and spatial relationships.

(2) The historic character of a property will be retained and preserved. The removal of distinctive materials or alteration of features, spaces and spatial relationships that characterize a property will be avoided.

(3) Each property will be recognized as a physical record of its time, place and use. Changes that create a false sense of historical development, such as adding conjectural features or elements from other historic properties, will not be undertaken.

(4) Changes to a property that have acquired historic significance in their own right will be retained and preserved.

(5) Distinctive materials, features, finishes and construction techniques or examples of craftsmanship that characterize a property will be preserved.

(6) Deteriorated historic features will be repaired rather than replaced. Where the severity of deterioration requires replacement of a distinctive feature, the new feature will match the old in design, color, texture and, where possible, materials. Replacement of missing features will be substantiated by documentary and physical evidence.

(7) Chemical or physical treatments, if appropriate, will be undertaken using the gentlest means possible. Treatments that cause damage to historic materials will not be used.

(8) Archeological resources will be protected and preserved in place. If such resources must be disturbed, mitigation measures will be undertaken.

(9) New additions, exterior alterations or related new construction will not destroy historic materials, features and spatial relationships that characterize the property. The new work will be differentiated from the old and will be compatible with the historic materials, features, size, scale and proportion, and massing to protect the integrity of the property and its environment.

(10) New additions and adjacent or related new construction will be undertaken in such a manner that, if removed in the future, the essential form and integrity of the historic property and its environment would be unimpaired.

(c)   *Restoration.*

(1) A property will be used as it was historically or be given a new use that interprets the property and its restoration period.

(2) Materials and features from the restoration period will be retained and preserved. The removal of materials or alteration of features, spaces and spatial relationships that characterize the period will not be undertaken.

(3) Each property will be recognized as a physical record of its time, place and use. Work needed to stabilize, consolidate and conserve materials and features from the restoration period will be physically and visually compatible, identifiable upon close inspection and properly documented for future research.

(4) Materials, features, spaces and finishes that characterize other historical periods will be documented prior to their alteration or removal.

(5) Distinctive materials, features, finishes and construction techniques or examples of craftsmanship that characterize the restoration period will be preserved.

(6) Deteriorated features from the restoration period will be repaired rather than replaced. Where the severity of deterioration requires replacement of a distinctive feature, the new feature will match the old in design, color, texture and, where possible, materials.

(7) Replacement of missing features from the restoration period will be substantiated by documentary and physical evidence. A false sense of history will not be created by adding conjectural features, features from other properties, or by combining features that never existed together historically.

(8) Chemical or physical treatments, if appropriate, will be undertaken using the gentlest means possible. Treatments that cause damage to historic materials will not be used.

(9) Archeological resources affected by a project will be protected and preserved in place. If such resources must be disturbed, mitigation measures will be undertaken.

(10) Designs that were never executed historically will not be constructed.

(d)   *Reconstruction.*
  (1) Reconstruction will be used to depict vanished or non-surviving portions of a property when documentary and physical evidence is available to permit accurate reconstruction with minimal conjecture and such reconstruction is essential to the public understanding of the property.
  (2) Reconstruction of a landscape, building, structure or object in its historic location will be preceded by a thorough archeological investigation to identify and evaluate those features and artifacts that are essential to an accurate reconstruction. If such resources must be disturbed, mitigation measures will be undertaken.
  (3) Reconstruction will include measures to preserve any remaining historic materials, features, and spatial relationships.
  (4) Reconstruction will be based on the accurate duplication of historic features and elements substantiated by documentary or physical evidence rather than on conjectural designs or the availability of different features from other historic properties. A reconstructed property will re-create the appearance of the non-surviving historic property in materials, design, color and texture.
  (5) A reconstruction will be clearly identified as a contemporary re-creation.
  (6) Designs that were never executed historically will not be constructed.

National Historic Register of Historic Places Standards

What Are the Criteria for Listing?

The National Register's standards for evaluating the significance of properties were developed to recognize the accomplishments of all peoples who have made a significant contribution to our country's history and heritage. The criteria are designed to guide State and local governments, Federal agencies, and others in evaluating potential entries in the National Register.

Criteria for Evaluation
The quality of significance in American history, architecture, archeology, engineering, and culture is present in districts, sites, buildings, structures, and objects that possess integrity of location, design, setting, materials, workmanship, feeling, and association, and:

  A. That are associated with events that have made a significant contribution to the broad patterns of our history; or
  B. That are associated with the lives of significant persons in our past; or

    C. That embody the distinctive characteristics of a type, period, or method of construction, or that represent the work of a master, or that possess high artistic values, or that represent a significant and distinguishable entity whose components may lack individual distinction; or

    D. That have yielded or may be likely to yield information important in history or prehistory.

Criteria Considerations

Ordinarily cemeteries, birthplaces, graves of historical figures, properties owned by religious institutions or used for religious purposes, structures that have been moved from their original locations, reconstructed historic buildings, properties primarily commemorative in nature, and properties that have achieved significance within the past 50 years shall not be considered eligible for the National Register. However, such properties will qualify if they are integral parts of districts that do meet the criteria or if they fall within the following categories:

    a. A religious property deriving primary significance from architectural or artistic distinction or historical importance; or

    b. A building or structure removed from its original location but which is primarily significant for architectural value, or which is the surviving structure most importantly associated with a historic person or event; or

    c. A birthplace or grave of a historical figure of outstanding importance if there is no appropriate site or building associated with his or her productive life; or

    d. A cemetery that derives its primary importance from graves of persons of transcendent importance, from age, from distinctive design features, or from association with historic events; or

    e. A reconstructed building when accurately executed in a suitable environment and presented in a dignified manner as part of a restoration master plan, and when no other building or structure with the same association has survived; or

    f. A property primarily commemorative in intent if design, age, tradition, or symbolic value has invested it with its own exceptional significance; or

    g. A property achieving significance within the past 50 years if it is of exceptional importance.

SOURCES: Code of Federal Regulations, Title 36, Chapter I, Part 68, from the Electronic Code of Federal Regulations (e-CFR), June 14, 2002: www.access.gpo.gov/nara/cfr/cfrhtml_00/Title_36/36cfr68_00.html; Criteria for Evaluation from the National Park Service, National Register of Historic Places Web site, June 18, 2002: www.cr.nps.gov/nr/listing.htm.

References

Alva Balderrama, Alejandro
1989 Earthquake damage to historic masonry structures. In *Conservation of Building and Decorative Stone*, vol. 2, ed. John Ashurst and Francis G. Dimes. London: Butterworth-Heinemann.

Araoz, Gustavo Jr., and Brian L. Schmuecker
1987 Discrepancies between U.S. national preservation policy and the Charter of Venice. In *Symposium Papers, ICOMOS General Assembly: Old Culture in New Worlds*, vol. 2, subtheme 4. Washington, D.C.: ICOMOS.

Bowman, J. N.
1951 *Adobe Houses in the San Francisco Bay Region*. San Francisco: California Division of Mines.

Brandi, Cesare
1977 *Principles for a Theory of Restoration*. Trans. Annalisa D'Amico. Rome: ICCROM.

California's State Historical Building Safety Code Board
1999 *State Historical Building Code (SHBC): Part 8, Title 24, California Code of Regulations, revised June 1998*. Sacramento, Calif.: State Historical Building Safety Code Board.

Craigo, Steade
1992 Conversation with Edna Kimbro.

Delong, David, ed.
1980 *Historic American Buildings: California*. 3 vols. New York and London: Garland Publishing.

EERI
1994 *Expected Seismic Performance of Buildings*. Oakland, Calif.: Earthquake Engineering Research Institute.

Feilden, Bernard
1988 Conservation of historic buildings between two earthquakes. In *Proceedings of the First International Seminar on Modern Principles in Conservation and Restoration of Urban and Rural Cultural Heritage in Seismic-Prone Regions*, 31–38. Skopje, Yugoslavia: Institute of Earthquake Engineering and Engineering Seismology (IZIIS).

Gavrilovic, Predrag, V. Sendova, Lj. Taskov, L. Krstevska, W. S. Ginell, and E. L. Tolles
1996 Shaking Table Tests of Adobe Structures. Report IZIIS 96-36. Skopje, Republic of Macedonia.

Ginell, William S., and E. Leroy Tolles
1999 Preserving safety and history: The Getty Seismic Adobe Project at work. *Conservation* 14, no. 1: 12–14.

2000 Seismic stabilization of historic adobe structures. *Journal of the American Institute for Conservation* 39, no. 1: 147–63.

Ginell, William S., E. Leroy Tolles, P. Gavrilovic, L. Krstevska, V. Sendova, and L. Taskov
2001 *Getty Seismic Adobe Project: Report of Third Year Activities: Shaking Table Tests of Large Scale Adobe Structures*. Los Angeles: Getty Conservation Institute.

Green, Melvyn
1990 Structural evaluation of Colton Hall. Report. Melvyn Green and Associates, Torrance, Calif.

Hamburger, Ronald O., Anthony B. Court, and Jeffrey R. Soulages
1995 Vision 2000: A framework for performance based engineering of buildings. In *Proceedings of the 64th Annual Convention of the Structural Engineers Association of California.* Whittier, Calif.: Structural Engineers Association of California.

Harthorn, Roy
1998 *Temporary Shoring and Stabilization of Earthquake-Damaged Historic Buildings.* Santa Barbara, Calif.: Roy Harthorn.

Historic American Buildings Survey (HABS)
1933 Library of Congress, Prints and Photographs Division. Washington, D.C.

ICOMOS
1999 ICOMOS charters and other international documents. *US/ICOMOS Scientific Journal* 1, no. 1. (Reviews of conservation and restoration charters and guidelines issued between 1904 and 1999.)

Imwalle, Michael
1992 Conversation with Edna Kimbro.

Imwalle, Michael, and M. Wayne Donaldson
1992 Conversation with Edna Kimbro.

Index of American Design
1943 Wall decoration over doorway. Archives of the National Gallery of Art, Washington, D.C., no. 1943.8.5941.

Jandl, H. Ward
1988 *Rehabilitating Interiors in Historic Buildings: Identifying and Preserving Character-Defining Elements.* Preservation Briefs 18. Washington, D.C.: U.S. Department of the Interior, National Park Service, Preservation Assistance Division.

Jokilehto, Jukka
1985 Authenticity in restoration principles and practices. Presentation, *APT Bulletin—The Journal of Preservation Technology* (Association for Preservation Technology) 12, nos. 3 and 4: 5–11.

1988 Modern principles of architectural conservation. In *Proceedings of the First International Seminar on Modern Principles in Conservation and Restoration of Urban and Rural Cultural Heritage in Seismic-Prone Regions,* 1–8. Skopje, Yugoslavia: Institute of Earthquake Engineering and Engineering Seismology (IZIIS).

Magalousis, Nicholas
1994 Conversation with Edna Kimbro.

Maish, James H.
1992 Mission accomplished: A landmark is rehabbed and restored. *Historic Preservation News* (National Trust for Historic Preservation) November.

Maroevic, Ivo
1988 Material structure and authenticity. In *Proceedings of the First International Seminar on Modern Principles in Conservation and Restoration of Urban and Rural Cultural Heritage in Seismic-Prone Regions,* 9–16. Skopje, Yugoslavia: Institute of Earthquake Engineering and Engineering Seismology (IZIIS).

Marquis-Kyle, Peter, and Meredith Walker
1992 *The Illustrated Burra Charter: Making Good Decisions about the Care of Important Places.* Sydney: Australia ICOMOS.

Monterey Historic Preservation Commission
1992 Minutes of meeting. City of Monterey, California, Planning Department. Manuscript on file.

Murphy, Donald
1992 Conversation with Edna Kimbro.

Nelson, Lee H.
n.d. *Architectural Character: Identifying the Visual Aspects of Historic Buildings as an Aid to Preserving Their Character.* Preservation Briefs 17. Washington, D.C.: U.S. Department of the Interior, National Park Service, Preservation Assistance Division.

Neuerburg, Norman
1977 Painting in the California missions. *American Art Review* IV, no. 1.

1987 *The Decoration of the California Missions.* Santa Barbara, Calif.: Bellerophon.

Oddy, Andrew, and Sara Carroll, eds.
1999 *Reversibility—Does It Exist?* British Museum Occasional Paper no. 135. London: British Museum.

Park, Sharon, Kay Weeks, Lauren Meier, Tim Buehner, and J. Ward Jandl
1991 *Preserving the Past and Making It Accessible to Everyone: How Easy a Task?* CRM Supplement. Washington, D.C.: U.S. Department of the Interior, National Park Service, Cultural Resources Preservation Assistance.

Riegl, Alois
1964 *Proceedings of the Second International Congress of Historic Monuments: Decisions and Resolutions (The Venice Charter).* Venice, Italy.

1982 The modern cult of monuments: Its character and its origin. Trans. Kurt Forster. *Oppositions 25: A Journal for Ideas and Criticism in Architecture* 25 (fall): 21–51.

Ruskin, John
1851–53 *The Stones of Venice.* London: Smith, Elder, and Co.

Seismic Safety Commission
1987 *Guidebook to Identify and Mitigate Seismic Hazards in Buildings.* Seismic Safety Commission Report no. SSC 87-03. Sacramento, Calif.: Seismic Safety Commission.

2000 *Status of the Unreinforced Masonry Building Law.* 2000 Annual Report to the Legislature. Seismic Safety Commission Report no. SSC 00-02. Sacramento, Calif.: Seismic Safety Commission.

Smith, Wallace E.
1977 *This Land Was Ours: The Del Valles and Camulos.* Ed. Grant W. Heil. Ventura, Calif.: Ventura Historical Society.

Structural Engineers Association of California
1999 *Recommended Lateral Force Requirements and Commentary.* Whittier, Calif.: Structural Engineers Association of California.

Thiel, Charles Jr., E. Leroy Tolles, Edna E. Kimbro, Frederick A. Webster, and William S. Ginell
1991 GSAP: Getty Conservation Institute guidelines for seismic strengthening of adobe project: Report of first year activities. Report. Getty Conservation Institute, Los Angeles.

Thomas, David Hurst, ed.
1991 Harvesting Ramona's garden: Life in California's mythical mission past. In *Colombian Consequences: The Spanish Borderlands in Pan-American Perspective,* vol. 3. Washington, D.C.: Smithsonian Institution Press.

Thomasen, S. E., and Carolyn L. Searls
1991 Seismic retrofit of historic structures in California. In *Structural Repair and Maintenance of Historical Buildings II: Proceedings of the 2nd International Conference, Seville, Spain,* vol. 2, 45–51. Southampton, U.K.: Computational Mechanics Publications.

Tolles, E. Leroy, Edna E. Kimbro, Charles C. Thiel Jr., Frederick A. Webster, and William S. Ginell
1993 GSAP: Getty Conservation Institute guidelines for seismic strengthening of adobe project: Report of second year activities. Report. Getty Conservation Institute, Los Angeles.

Tolles, E. Leroy, Edna E. Kimbro, Frederick A. Webster, and Anthony Crosby
1996 *Survey of Damage to Historic Adobe Buildings after the January 1994, Northridge Earthquake.* GCI Scientific Program Reports. Los Angeles: Getty Conservation Institute.

Tolles, E. Leroy, Edna E. Kimbro, Frederick A. Webster, and William S. Ginell
2000 *Seismic Stabilization of Historic Adobe Structures: Final Report of the Getty Seismic Adobe Project.* GCI Scientific Program Reports. Los Angeles: Getty Conservation Institute.

Tolles, E. Leroy, and Helmut Krawinkler
1990 Seismic studies on small-scale models of adobe houses. Ph.D. diss., John A. Blume Earthquake Engineering Center, Department of Civil Engineering, Stanford University, California.

Uniform Building Code (UBC): Modern Criteria for Seismic Design and Construction
1997 Whittier, Calif.: International Conference of Building Officials.

Vargas-Neumann, Julio
1984 *Earthquakes and Earthen Structures.* Report. ICCROM, Rome.

Viollet-le-Duc, E. E.
[1858–72] 1959 *Discourses on Architecture.* London: G. Allen and Unwin.

Watkins, Malcolm C.
1973 Observations and opinions about the Boronda Adobe. Manuscript on file at the Monterey County Historical Association, Salinas, California.

Additional Reading

Allen, D., G. Sanchez, and J. Hill. The effects of the Loma Prieta earthquake on the seismically retrofitted Santa Cruz Mission Adobe. In *The Seismic Retrofit of Historic Buildings Conference Workbook*. San Francisco: Western Chapter of the Association for Preservation Technology, 1991.

Ambraseys, N. N. An earthquake engineering viewpoint of the Skopje earthquake, 26th July, 1963. In *Proceedings of the Third World Conference on Earthquake Engineering*, vol. 3, S-22–S-38. Wellington: New Zealand National Committee on Earthquake Engineering, 1965.

Arango, I., et al. Adobe-type dwellings: A method to optimize their replacement. In *Proceedings, Ninth World Conference on Earthquake Engineering*, vol. 7, paper 13-1-7, 563–68. Tokyo: 9WCEE Organizing Committee, Japan Association for Earthquake Disaster Prevention, 1989.

Architectural Resources Group. *An Assessment of Damage Caused to Historical Resources by the Loma Prieta Earthquake*. San Francisco: National Trust for Historic Preservation, 1990.

Association for Preservation Technology. Historic structure reports: Special issue. *APT Bulletin: The Journal of Preservation Technology* 28, no. 1. Eleven papers on the use of historic structure reports, 1997.

Attar, Ghassan. Authenticity vs. stability: The conservation engineer's dilemma. *APT Bulletin: The Journal of Preservation Technology* 23, no. 1 (1991).

Ayres, James E., comp. *The Archaeology of Spanish and Mexican Colonialism in the American Southwest, Colombian Quincentenary Series*. Guides to the Archaeological Literature of the Immigrant Experience in America, no. 3. Ann Arbor: Society for Historical Archaeology, 1995.

Bariola, J., and M. A. Sozen. Seismic tests of adobe walls. *Earthquake Spectra* 6, no. 1 (1990): 37–56.

Bates, F. L., and C. D. Killian. Changes in housing in Guatemala following the 1976 earthquake, with special reference to earthen structures and how they are perceived by disaster victims. *Disasters* 6, no. 2 (1982): 92–100.

Blumenthal, Sara K., comp. *Federal Historic Preservation Laws*. Revised by Emogene A. Bevitt. Washington, D.C.: U.S. Department of the Interior, National Park Service, Cultural Resources Programs, 1993.

The California Missions. Menlo Park, Calif.: Sunset Publishing, 1979.

California Native American Heritage Commission. *A Professional Guide for the Preservation and Protection of Native American Remains and Associated Grave Goods*. Sacramento: California Native American Heritage Commission, 1991.

Cao, S. K., and L. Zhao. *The Earthquake Resistant Characteristics of the Raw Earth Buildings in Ningxia*. Yinchuan, China: Ningxia Industrial Design Institute, 1981.

Castano, J. C., et al. The possible influence of soils conditions on earthquake effects: A case study. In *Proceedings, Eleventh World Conference on Earthquake Engineering,* disc 2, paper no. 1068. Oxford: Pergamon, Elsevier Science, 1996.

Cox, Rachel S. *Controlling Disaster: Earthquake Hazard Reduction for Historic Buildings.* Information Series no. 61. Washington, D.C.: Western Regional Office, National Trust for Historic Preservation, 1992.

Delahanty, Randolph, and E. Andrew McKinney. *Preserving the West.* New York: Pantheon Press, 1985.

Dudley, E. Disaster mitigation: Strong houses or strong institutions? *Disasters* 12, no. 2 (1988): 111–21.

Earthen Building Technologies. *Workshop on the Seismic Retrofit of Historic Adobe Buildings.* Pasadena, Calif.: Earthen Building Technologies, 1995.

Erdik, M. O., ed. *Middle East and Mediterranean Regional Conference on Earthen and Low-Strength Masonry Buildings in Seismic Areas, Ankara, Turkey.* Dallas: Intertect, 1986.

Feilden, Bernard M. *Between Two Earthquakes: Cultural Property in Seismic Zones.* Marina del Rey, Calif.: Getty Conservation Institute; Rome: ICCROM, 1987.

Freeman, Allen. Adobe duo. *Historic Preservation* 43, no. 4, 1991.

Gere, James M., and Haresh C. Shah. *Terra Non Firma: Understanding and Preparing for Earthquakes.* The Portable Stanford. Stanford, Calif.: Stanford Alumni Association, 1984.

Getty Conservation Institute. Workshop on the seismic retrofit of historic adobe buildings. Report of conference held at the J. Paul Getty Museum, Malibu, California. Los Angeles: Getty Conservation Institute, 1995.

Glass, R. I., et al. Earthquake injuries related to housing in a Guatemalan village— aseismic construction techniques may diminish the toll of deaths and serious injuries. *Science* 197, no. 4304 (1977): 638–43.

Grieff, Constance. *The Historic Property Owner's Handbook.* Washington, D.C.: Preservation Press, National Trust for Historic Preservation, 1977.

Guidelines for the Evaluation of Historic Unreinforced Brick Masonry Buildings in Earthquake Hazard Zones (ABK Methodology). Sacramento: California State Department of Parks and Recreation, 1986.

H. J. Degenkolb Associates. *Balancing Historic Preservation and Seismic Safety.* San Francisco: H. J. Degenkolb Associates, 1992.

Hovey, Lonnie J. Evolving preservation standards and strategies for the octagon: Reemphasizing the significance of structural fabric. *Historic Preservation Forum* 5, no. 2 (March/April 1991).

Jester, Thomas C., and Sharon C. Park. *Making Historic Properties Accessible.* NPS Preservation Briefs 32. Washington, D.C.: U.S. Department of the Interior, National Park Service, Preservation Assistance Division, 1993.

Jones, Barclay, ed. *Cultural Heritage in Peril from Natural Disasters; Protecting Historic Architecture and Museum Collections from Earthquakes and Other Natural Hazards.* Washington, D.C.: Architectural Research Centers Consortium, 1984.

Kirker, Harold. California architecture and its relation to contemporary trends in Europe and America. *California Historical Quarterly* (winter 1978).

Langenbach, Randolph. Bricks, mortar, and earthquakes: Historic preservation vs. earthquake safety. *APT Bulletin: The Journal of Preservation Technology* 21, no. 3/4 (1989).

Look, David W. Preservation and seismic retrofit of historic resources: NPS technical assistance in the Middle East. *Cultural Resource Management* 16, no. 7 (1993).

Look, David W. *The Seismic Retrofit of Historic Buildings*. NPS Preservation Briefs 41. Washington, D.C.: U.S. Department of the Interior, National Park Service, Preservation Assistance Division, 1997.

Mahmood, H. Damages in masonry structures and counter measures. *Bulletin of the International Institute of Seismology and Earthquake Engineering* 26 (1992): 177–84.

May, G. W., and Frederick C. Cuny, eds. International workshop on earthen buildings in seismic areas. In *Proceedings of International Workshop, University of New Mexico, Albuquerque, May 24–28*. Albuquerque: Univ. of New Mexico Press, 1981.

Merritt, John. *History at Risk*. Oakland, Calif.: California Preservation Foundation, 1990.

Meli, Roberto, Oscar Hernandez, and M. Dadilla. Strengthening of adobe houses for seismic actions. In *Proceedings of the Seventh World Conference on Earthquake Engineering*, vol. 4, 465–72. Istanbul: Turkish National Committee on Earthquake Engineering, 1980.

Neumann, Julio Vargas. Earthquake resistant rammed earth (*tapial*) buildings. In *Terra 93: Proceedings of the Seventh International Conference on the Study and Conservation of Earthen Architecture, Silves, Portugal, October 1993*, edited by Margarida Alcada, 503–8. Rome: ICCROM, 1993.

Philippot, Paul. Historic preservation: Philosophy, criteria, guidelines. In *Preservation and Conservation: Principles and Practices*. Washington, D.C.: Preservation Press, 1976.

Qamaruddin, M., and B. Chandra. Behaviour of unreinforced masonry buildings subjected to earthquakes. *The Masonry Society Journal* 9, no. 2 (1991): 47–55.

Razani, R. Investigation of lateral resistance of masonry and adobe structures by means of a tilting table. In *Proceedings, Sixth World Conference on Earthquake Engineering, Meerut, India*, edited by Sarita Prakashan, vol. 2, 2130–31. Roorkee, India: Society of Earthquake Technology, 1977.

———. Seismic protection of unreinforced masonry and adobe low-cost housing in less developed countries: Policy issues and design criteria. *Disasters* 2, no. 2/3 (1978): 137–47. Roorkee, India.

Roselund, Nels. Repair of cracked adobe walls by injection of modified mud. In *Adobe 90: Proceedings of the Sixth International Conference on the Conservation of Earthen Architecture, Las Cruces, New Mexico, USA, October 1990*, ed. Neville Agnew, Michael Taylor, and Alejandro Alva Baderama, 336–41. Los Angeles: Getty Conservation Institute, 1990.

Scawthorn, Charles. Relative benefits of alternative strengthening methods for low strength masonry buildings. In *Proceedings of the Third National Conference on Earthquake Engineering, Charleston, South Carolina*. Oakland, Calif.: Earthquake Engineering Research Institute, 1986.

Schuetz-Miller, Mardith K. *Architectural Practice in Mexico City: A Manual for Journeyman Architects of the Eighteenth Century*. Tucson: Univ. of Arizona Press, 1987.

Secretary of the Interior. *The Secretary of the Interior's Standards for the Treatment of Historic Properties with Guidelines for Preserving, Rehabilitating, Restoring, and Reconstructing Historic Buildings*. Washington, D.C.: U.S. Department of the Interior, National Park Service, Preservation Assistance Division, 1995.

Seismic Safety Commission. *Architectural Practice and Earthquake Hazards: A Report of the Committee on the Architect's Role in Earthquake Hazard Mitigation*. Publication no. SSC 91-10. Sacramento, Calif.: Seismic Safety Commission, 1991.

———. *Earthquake Risk Management Tools for Decision Makers: A Guide for Decision Makers*. Publication no. SSC 99-06. Sacramento, Calif.: Seismic Safety Commission, 1999.

Spence, R. J. S., et al. Correlation of ground motion with building damage: The definition of a new damage-based seismic intensity scale. In *Proceedings of the Tenth World Conference on Earthquake Engineering, Madrid, Spain.*, vol. 1, 551–56. Rotterdam: A. A. Balkema, 1992.

Spennemann, Dirk H. R., and David W. Look, eds. *Disaster Management Programs for Historic Sites*. Washington, D.C.: National Park Service, Association for Preservation Technology, Western Chapter; Albury, Australia: Johnstone Centre, Charles Sturt University, 1998. The following chapters are of special interest:

- Donaldson, Milford Wayne. Conserving the historic fabric: A volunteer disaster worker's perspective; The first ten days: Emergency response and protection strategies for the preservation of historic structures.

- Kariotis, John. The tendency to demolish repairable structures in the name of life safety.

- Langenbach, Randolph. Architectural issues in the seismic rehabilitation of masonry buildings.

- Mackensen, Robert. Cultural heritage management and California's State Historical Building Code.

Webster, Frederick A., and J. D. Gunn. Seismic retrofit techniques for historic and older adobes in California. In *Terra 93: Proceedings of the Seventh International Conference on the Study and Conservation of Earthen Architecture, Silves, Portugal, October 1993,* edited by Margarida Alcada, 521–25. Rome: ICCROM, 1993.

Glossary

adobe   Outdoor, air-dried, unburned brick made from a clayey soil and usually mixed with straw or animal manure. The clay content of the soil ranges from 10% to 30%.

basal erosion   Coving-type deterioration at the base of an adobe wall.

bond beam   Wood or concrete beam that is attached to the top of a wall at the roof level and that encircles the perimeter of a building.

center-core rods   Steel or reinforced polymer rods inserted vertically in drilled holes in adobe walls. Set in a polyester, epoxy, or adobe grout, *center-core rods* are used to minimize the relative displacement of cracked wall sections.

corredor   Covered (roofed) exterior corridor or arcade called a portal or portico in New Mexico, also referred to as a veranda or porch.

cracked wall section   Portion of an adobe wall that is defined by a boundary of through-wall cracks.

diaphragm   A large, thin structural element, usually horizontal, that is structurally loaded in its plane. The *diaphragm* is usually an assemblage of elements that can include roof or floor sheathing, framing members to support the sheathing, and boundary or perimeter members.

flexural stresses   Stresses in an object that result from bending.

flexure   Bending.

foundation settlement   Downward movement of a foundation caused by subsidence or consolidation of the supporting ground.

freestanding walls   Walls, such as garden walls, that are supported only laterally at ground level and have no attached roof or floor framing.

ground motion   Lateral or vertical movement of the ground, as in an earthquake.

HABS   Historic American Buildings Survey. An ongoing federal documentation program of historic buildings in the United States, initiated as part of the Works Progress Administration (WPA) in the 1930s.

headers   Adobe blocks placed with the long direction perpendicular to the plane of the wall.

in plane   Deflections or forces that are parallel to the plane of a wall.

joist   Closely spaced horizontal beams (spaced at approximately 0.6 m [2 ft.] on center) that span an area, such as a floor or ceiling.

ladrillo   A flat, baked terra-cotta tile or brick.

lintel	A horizontal structural member that spans the opening over a window or a door in a wall and usually carries the weight of the wall above the opening. In historic adobe buildings, *lintels* are usually made of wood.
load-bearing	Building elements, such as walls, that carry vertical loads from floors or roofs.
non-load-bearing	Building elements, such as walls, that do not carry vertical loads from floors or roofs.
out of plane	Deflections or forces that are perpendicular to the plane of a wall.
overturning	Collapse of a wall caused by rotation of the wall about its base.
rafters	Parallel, sloping timbers or beams that give form and support to a roof.
shear forces	Typically, in adobe walls, forces that occur in the plane of the wall and cause diagonal cracking. *Shear forces* can also be developed out of plane in a wall and are caused by forces that produce an opposite but parallel sliding motion across an interface in the wall.
slenderness ratio (S_L)	The ratio of the height of a wall to its thickness. Herein, the slenderness ratio for thick walls is taken as $S_L < 6$; for moderately thick walls, $S_L = 6\text{–}8$; and for thin walls, $S_L > 8$.
slumping	Bulging at the base of an adobe wall caused by moisture intrusion, resulting in an increase in the adobe's plasticity and loss of the material's strength.
straps	Flexible cords, cables, or flat woven ribbons that encircle the walls, are used to minimize relative displacements, and hold cracked adobe blocks in the plane of the wall during seismic shaking. These may be made of nylon, polypropylene, or other strong, stable polymer. Steel cables may also be used.
stretchers	Adobe blocks placed with the long direction parallel to the plane of the wall.
tapanco	An attic, loft, garret, or half-story of a building that is accessed by stairs or a ladder in the gable-end wall.
wet-dry cycles	Repeated cycles of water exposure and drying out that can lead to a loss of cohesion of the clay particles in the adobe. This usually results in a weakened material.

Cumulative Index to the
Getty Seismic Adobe Project Volumes

About the Authors

William S. Ginell is a materials scientist with extensive experience in industry. In 1943, after graduating from the Polytechnic Institute of Brooklyn with his bachelor's degree in chemistry, he became part of the secret research team at Columbia University working to develop the atomic bomb. After the war, he went on to receive his Ph.D. in physical chemistry from the University of Wisconsin, spent nine years at the Brookhaven National Laboratory on Long Island, New York, followed by twenty-six years working for aerospace firms in California. In 1984 he joined the Getty Conservation Institute as head of Materials Science. He is currently senior conservation research scientist at GCI and was project director of the Getty Seismic Adobe Project.

Edna E. Kimbro is an architectural conservator and historian, specializing in research and preservation of Spanish and Mexican colonial architecture and material culture of early California. She studied architectural history at the University of California, Santa Cruz. Through the 1980s she was involved in restoration of the Santa Cruz Mission Adobe for California State Parks. In 1989 she attended ICCROM in Rome and studied seismic protection of historic adobe buildings. In 1990 she became preservation specialist for the Getty Seismic Adobe Project. Currently, she is the Monterey (California) District historian for California State Parks and prepares historic structure reports for adobe buildings.

E. Leroy Tolles has worked on the seismic design, testing, and retrofit of adobe buildings since the early 1980s, specializing in the structural design and construction of earthen and wood buildings. He received his doctorate from Stanford University in 1989, where his work focused on the seismic design and testing of adobe houses in developing countries. He has led multidisciplinary teams to review earthquake damage after the 1985 earthquake in Mexico, the 1989 Loma Prieta earthquake, and the 1994 Northridge earthquake. He was principal investigator for the Getty Conservation Institute's Getty Seismic Adobe Project and has coauthored numerous publications on seismic engineering. He is principal for ELT & Associates, an engineering and architecture firm in Northern California.